Pedro Oria Iriarte

Double-slit experiments and decay probability in microwave billiards

Pedro Oria Iriarte

Double-slit experiments and decay probability in microwave billiards

Südwestdeutscher Verlag für Hochschulschriften

Imprint
Any brand names and product names mentioned in this book are subject to trademark, brand or patent protection and are trademarks or registered trademarks of their respective holders. The use of brand names, product names, common names, trade names, product descriptions etc. even without a particular marking in this work is in no way to be construed to mean that such names may be regarded as unrestricted in respect of trademark and brand protection legislation and could thus be used by anyone.

Publisher:
Südwestdeutscher Verlag für Hochschulschriften
is a trademark of
Dodo Books Indian Ocean Ltd., member of the OmniScriptum S.R.L Publishing group
str. A.Russo 15, of. 61, Chisinau-2068, Republic of Moldova Europe
Printed at: see last page
ISBN: 978-3-8381-2066-9

Zugl. / Approved by: Darmstadt, TU, Diss.,2010

Copyright © Pedro Oria Iriarte
Copyright © 2010 Dodo Books Indian Ocean Ltd., member of the OmniScriptum S.R.L Publishing group

Double-slit experiments and decay probability in microwave billiards

Vom Fachbereich Physiks
der Technischen Universität Darmstadt

zur Erlangung des Grades
eines Doktors der Naturwissenschaften
(Dr. rer. nat.)

genehmigte

D i s s e r t a t i o n

angefertigt von

MSc. Pedro Oria Iriarte
aus Pamplona (Spanien)

Darmstadt 2010
D 17

Referent: Professor Dr. rer. nat. Dr. h.c. mult. A. Richter
Korreferent: Professor Dr. rer. nat. J. Wambach

Tag der Einreichung: 10.02.2010
Tag der Prüfung: 26.04.2010

Abstract

The present doctoral thesis is concerned, on the one hand, with the investigation of the decay probability in a quantum billiard, and, on the other, the realization of the double-slit experiment in microwave billiards. For both purposes transmission elements of the scattering matrix are experimentally measured in microwave cavities with regular and chaotic dynamics. One or more pieces of the billiards' boundary are removed and the leaked outward field is investigated. The general aim is to distinguish the effect of the billiard dynamics on the temporal decay behaviour and on the interference patterns.

The decay probability in an open quantum billiard can be explored by means of the scattering matrix formalismus derived in the framework of nuclear physics. In a circular billiard, archetype of regular systems, the opening breaks the spatial symmetry causing a splitting of the degenerate states observable at low frequencies. One state of the doublet remains unaffected contributing to the long-time behavior inside the cavity (survival probability), while the short-lived one is strongly influenced by the opening and decays much faster. The case of chaotic dynamics is studied by means of a tilted stadium billiard. For sufficiently long times the decay probability shows an essentially nonexponential behaviour. Agreement with theoretical results derived by Harney, Dittes and Müller is obtained for a few of the open channels and intermediate times. For both, the circle and the stadium, the short time decay resembles the classical behaviour.

Furthermore, a double-slit experiment is performed with a wave packet starting inside a microwave billiard domain, both for regular and chaotic dynamics. Stationary and temporal patterns formed by the leaking of the wave packet are experimentally determined. The evolution of a directed initial state is constructed by means of the electromagnetic propagator. Also single-slit experiments are carried out and compared with the corresponding double-slit experiments.

The present work sheds light on the fundamental question of the decay of open quantum systems and provides plentiful information about the structure of the diffracted field from microwave billards with a double-slit on the boundary connecting classical properties with the corresponding quantum features in both cases.

Zusammenfassung

Die vorliegende Doktorarbeit beschäftigt sich, zum einen, mit der Untersuchung der Zerfallswahrscheinlichkeit in einem Quantenbillard und, zum anderen, mit der Durchführung des Doppeltspaltexperiments in Mikrowellenbillards. Für beide Zwecke werden Transmissionselemente der Streumatrix in Mikrowellenkavitäten mit regulärer und chaotischer Dynamik experimentell gemessen. Ein oder mehrere Stücke der Billiardberandung werden entfernt und das auslaufende Feld wird weitreichend betrachtet. Hauptziel ist, den Effekt der Billarddynamik auf das zeitliche Zerfallsverhalten und Interferenzmuster zu unterscheiden.

Die Zerfallswahrscheinlichkeit in einem offenen Quantenbillard kann mittels des Streumatrix-Formalismus aus der Kernphysik untersucht werden. In einem Kreisbillard, Archetyp eines regulären Systems, bricht die Öffnung die örtliche Symmetrie und bewirkt die Aufspaltung der entarteten Zustände. Diese ist beobachtbar für niedrige Anregungsfrequenzen. Ein Zustand des Dupplets bleibt unbeeinflusst und trägt zum Langzeitverhalten bei (Aufenthaltswahrscheinlichkeit), während der kurzlebige stark beeinflusst von der Öffnung ist und dementsprechend viel schneller zerfällt. Der Fall der chaotischen Dynamik wird mittels eines abgeschrägten Stadiumbilliards untersucht. Für ausreichend lange Zeiten zeigt die Zerfallswahrscheinlichkeit ein nicht exponentielles Verhalten. Übereinstimmung mit von Harney, Dittes and Müller entwickelten theoretischen Resultaten wird für eine kleine Anzahl von offenen Kanälen und mittlere Zeiten gefunden. Sowohl für das Kreis- als auch für das Stadiumbilliard nähert sich die Kurzzeitdynamik dem entsprechenden klassichen Verhalten.

Ein Doppeltspaltexperiment wird mit einem in einen Mikrowellenbillard regulärer bzw. chaotischer Dynamik startenden Anfangswellenpaket realisiert. Stationäre- und Zeitmuster vom gebeugten Wellenpacket werden experimentell gemessen. Die Entwicklung eines ausgerichteten Anfangszustandes wird mit Hilfe des elektromagnetischen Propagators rekonstruiert. Einzelspaltexperimente werden ebenfalls realisiert und mit den entsprechenden Doppeltspaltexperimenten verglichen.

Die vorliegende Arbeit wirft Licht auf die grundlegende Frage des Zerfalls eines offenen Quantensystems und liefert reichlich Information über die Struktur des gebeugten Feldes aus Mikrowellenbillards mit Doppelschlitzen am Rand. In beiden Fällen werden klassische Eigenschaften mit den von analogen Quantensystemen verknüpft.

Contents

1	**Introduction**	**1**
2	**Basic principles**	**4**
	2.1 Open classical billiards	4
	2.2 Quantum and microwave billiards	6
3	**Escape probabilities from classical billiards**	**8**
4	**Experiments with microwave billiards**	**10**
	4.1 Measurements at room temperature	10
	4.2 Measurements at superconducting conditions	14
5	**Survival probability in open quantum billiards**	**16**
	5.1 Description of open quantum systems	16
	5.1.1 Survival and decay probabilities	16
	5.1.2 Resonance widths	20
	5.2 Experiment	22
	5.2.1 Design of the resonators	22
	5.2.2 Measurements of intensity distributions of the open circle billiard	24
	5.2.3 Measurements of frequency spectra	28
	5.3 Analysis and interpretation	34
	5.3.1 Temporal decay behaviour of single resonances	34
	5.3.2 Temporal decay behaviour of the open circle billiard	36
	5.3.2.1 Long-time behaviour in the open circular billiard	38
	5.3.2.2 Short-time behaviour in the open circular billiard	42

		5.3.2.3	Length spectra	42
	5.3.3	\multicolumn{2}{l}{Temporal decay behaviour of the open stadium billiard . .}	45
		5.3.3.1	Long-time behaviour in the open stadium billiard	46
		5.3.3.2	Short-time behaviour in the open stadium billiard	49
5.4	\multicolumn{3}{l}{Conclusions . . . .}	51

6 Double-slit experiments with microwave billiards   52

- 6.1 Motivation 52
- 6.2 Experiment 55
- 6.3 Stationary wave patterns 60
- 6.4 Interference patterns in the time domain 65
  - 6.4.1 Time-dependent Schrödinger equation versus Helmholtz equation 65
  - 6.4.2 Emission from a single antenna in free space 66
  - 6.4.3 Emission from a single antenna in microwave billiards . . . 68
- 6.5 Excitation of the billiards with an elongated wave packet 74
  - 6.5.1 Preparation of an elongated wave packet 74
  - 6.5.2 Experiments with an elongated wave packet 77
- 6.6 Excitation of the billiards with a Gaussian wave packet 81
  - 6.6.1 Preparation of a Gaussian wave packet 81
  - 6.6.2 Experiments with a Gaussian wave packet in free space . . 83
  - 6.6.3 Experiments with a Gaussian wave packet in microwave billiards 83
- 6.7 Conclusions 88

7 Final considerations   90

1 Introduction

In 1887, the French mathematician Henri Poincaré studied the motion of more than two orbiting bodies in the solar system. His findings led to the development of a deterministic chaos theory [1, 2]. Small differences in the initial conditions of the bodies' motion entailed increasing exponential deviations as time proceeded. In the next 100 years, chaotic systems received a lot of attention from both the theoretical [3–7] and the experimental point of view [8–10]. Billiards have played a decisive role in the understanding of chaotic behaviour. These are bounded domains in which a point-like particle is confined by infinitely high hard walls and moves freely [11]. If the particle collides with the walls of the billiard it is specularly reflected. Billiards are the object of vivid investigation since the dynamics of the point-like particle depends only on the shape of the billiard, and may be regular, chaotic or mixed.

The study of quantized systems whose classical analoga exhibit chaotic features is covered by the field of so called quantum chaos [5, 7]. It attempts to build a connection between the theory of quantum mechanics and the classical dynamics of chaotic systems. In quantum mechanics, the definition of chaos as sensitive dependence on initial conditions turns out to be futile since the concept of exponential separation of neighboring trajectories loses its meaning due to the uncertainty relations. Besides, quantum motion is dynamically stable, i.e. initial errors propagate only linearly in time according to the Schrödinger equation. Linear instability is a typical feature of classical integrable systems and this contrasts the exponential instability characterizing classical chaotic systems. Therefore, it appears that quantum motion always exhibits the characteristic features of classically integrable, regular motion which is just the opposite of dynamical chaos. This apparently paradoxical situation can be resolved by introducing a time scale below which quantum mechanics mimics classical dynamics. This scale is called the Heisenberg time t_H and is defined by the inverse of the mean spacing between consecutive levels of a quantum system. The dynamics of quantum systems before and after t_H will be discussed in the course of the present doctoral thesis.

The theoretical understanding of quantum chaos rests mainly upon two pillars. On the one hand, the so called periodic orbit theory has been developed in the last 50 years [12]. The crucial result of this theory is the trace formula which constitutes a semiclassical quantization connecting the level density of quantum states with the periodic orbits of the corresponding classical system. On the other hand, Bohigas, Giannoni and Schmit predicted a universal relationship between the statistical properties of spectra of quantum mechanical systems with chaotic classical counterpart and random matrix theory [13]. Some mathematical aspects and theoretical foundations about billiards and the discipline of quantum chaos are given in Sec. 2.

In contrast to closed billiards, open classical billiards are characterized by an opening in the boundary through which particles are allowed to escape. Numerical simulations presented in Sec. 3 show the escape probabilities in open classical billiards with regular and chaotic dynamics.

Due to the equivalence between the time-independent Schrödinger equation and the scalar Helmholtz equation, the motion of a quantum particle can be experimentally simulated by electromagnetic waves inside two-dimensional microwave billiards [14]. Microwave billiards are flat resonators into which microwave power is coupled through dipolar antennas. The details are given in Sec. 2.2. In the last 20 years experiments using brass microwave cavities have been performed in several laboratories. Since 1992 the usage of superconducting cavities (made from niobium or lead-plated copper) at the Institute of Nuclear Physics in Darmstadt allows most precise studies of spectral properties of quantum chaotic systems [15, 16]. In contrast to measurements at normal conducting conditions, the ones at superconducting conditions are characterized by very small ohmic losses at the cavity walls.

Experimentally, disturbing a system is unavoidable to extract information from it. In measurements on microwave cavities, the antennas act as scattering channels and represent a coupling to the outside world. Absorption at the cavity walls can be modelled by a large number of additional open channels. Thus, the microwave cavity is an open scattering system. Besides, billiards can also be opened through small holes on the boundary and interchange energy with the exterior [17]. In the present doctoral thesis, the temporal decay behaviour and the structure of the outgoing waves is investigated for microwave billiards with openings in the walls. The aim is to distinguish between effects caused by regular and chaotic dynamics. The results are summarized in sections 5 and 6.

The study of time-dependent manifestations in the field of quantum chaos has aroused much interest as reflected in a large number of publications (see [18–20] and references therein). In particular, the quantity defined as survival probability (also denoted in the literature as return probability [21] or norm-leakage decay function [22]) is the focus of investigations in several works and it has been recently explored numerically [21] as well as experimentally in dielectric cavities [23] and in atom-optic billiards [24, 25]. The central question has been: what is the probability that an initial set of particles has not left the system until time t. Another closely related issue is the field of quantum transport which has an ample spectrum of applications in mesoscopic physics [26–31]. Semiclassical techniques are usually employed to describe such physical problems [11, 32–34]. Numerical results concerning the temporal escape law of wave packets in quantum billiards have been published in [35]. The authors found that the quantum-mechanical dynamics starts to deviate from the dynamics of its classical counterpart at a time scale t near t_H. For longer times algebraic decay laws are found in [35] both for regular and chaotic systems. In Sec. 5, the time decay behaviour of the current leaking out of quantum billiards is experimentally determined. The outgoing current is defined as the time derivative of the above mentioned survival probability and can be linked to quantities measured in the experiments through the theory of open quantum systems. In the present work a large variety of decay behaviour is found which can be related to resonance widths. As will be shown the realization of the experiments at superconducting conditions is of great importance for the distinguishability between decay of regular and chaotic billiards. Cavities with different opening sizes are investigated. The experiments realised in this doctoral thesis thus shed some light on the fundamental question of the decay of quantum systems.

At the beginning of the 1800s, Thomas Young performed for the first time an experiment which still attracts large interest in physics: the interference of light beams passing through a double slit [36]. This experiment and its outcome has played a profound role in the development of optics and quantum mechanics and has been used since as a paradigm to unveil the wave nature of a number of physical entitites, in particular single electrons [37, 38], neutrons [39], atoms [40, 41] and molecules [42]. In Sec. 6, single and double slit experiments with microwave billiards are presented. Interference patterns are analyzed by studying the temporal and spatial distribution of the electromagnetic field close to the slits in the experiments. The aim of this experimental investigation originates from a numerical simulation of a double-slit experiment in which a Gaussian wave packet initially confined inside a billiard domain with two slits in the boundary propagates [43]. The simulation revealed a fundamental distinction between diffraction patterns from billiards with regular and chaotic dynamics, at least for certain initial conditions.

In this thesis, measurements to verify (or falsify) the predictions of Ref. [43] have been performed. In order to choose a similar initial state as in the simulation of [43], the construction of a directed electromagnetic wave packet is achieved. A connection between classical periodic orbits, the time the wave packet spends inside the billiard before leaving via the slits and the initial direction of the wave packet is illustrated. This connection is clearly obtained if the wave packet sticks on some periodic orbits. Nevertheless, as shown in [43], only the choice of certain initial directions yield a clear distinction between patterns from regular and chaotic cavities.

2 Basic principles

2.1 Open classical billiards

The frictionless motion of a particle with mass m and momentum $\vec{p}$ on a plane billiard table bounded by hard walls represents the simplest situation for demonstrating regular and chaotic dynamics. The numerical treatment, as opposed to other systems, does not require the numerical integration of differential equations. The particles move along straight trajectories until they are reflected at the wall according to the rules of specular reflection. At a colllision with the wall the tangential component with respect to the boundary of the particle's momentum is not modified whereas the normal component is inverted. In an open classical billiard particles can escape through an opening of size Δ in the boundary (see Fig. 2.1).

A billiard is called regular if the trajectories of two particles diverge after an initial perturbation at most linearly in time. Examples are billiards with a

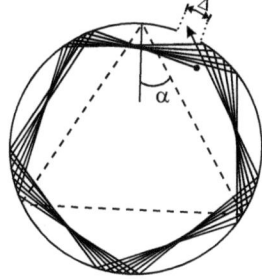 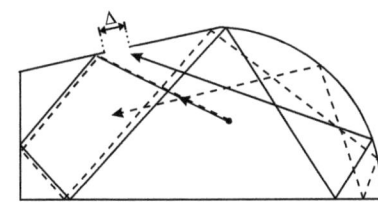

Fig. 2.1: Examples of orbits in open classical billiards. In the circle billiard (left panel), archetyp of regular systems, a triangular periodic orbit (dashed) and a quasi-periodic one (solid) are shown. In the tilted stadium (right panel), two trajectories with infinitesimally differing initial conditions deviate strongly after a few bounces at the boundary.

circular or rectangular shape which in particular are investigated in the present doctoral thesis. The circular billiard is a rotationally symmetric system and the dynamics is integrable, the constants of motion being the energy and the angular momentum. If the reflection angle at the boundary equals $\alpha = 2\pi m/n$, where m and n are integers with no common divisor, the particle moves along a closed periodic curve in coordinate space (see the dashed orbit in the left panel of Fig. 2.1). On the contrary, if α is an irrational multiple of π, the particle does not return to its original position and it eventually hits all points along the boundary (see the solid orbit in the left picture of Fig. 2.1). Such orbits are called quasiperiodic. In an open circle billiard they determine the long-time behaviour of the survival probability before the particle escapes.

The motion of a point-like particle in a chaotic billiard is unpredictable after several bounces. An example (see right picture of Fig. 2.1) for such a chaotic system is the tilted variant of a desymmetrized Bunimovich stadium [44]. A tilt of the upper segment assures the suppression of marginally stable families of periodic orbits bouncing between the upper and lower sides of a desymmetrized Bunimovich stadium (with the shape of a rectangle attached to a quarter circle), the so called bouncing ball orbits [45]. Consequently, the rectangle is replaced by a trapezoid. In a stadium billiard slight changes in the initial position and/or momentum of the particle lead to large deviations. After just a few collisions two initially closely lying and nearly parallel trajectories will move apart with an exponential rate of separation. This lack of predictability and instability are characteristic features of the dynamics in chaotic billiards and are quantitatively characterised by a Lyapunov exponent greater than zero[1]. In the right panel of Fig. 2.1, two orbits in the tilted stadium are represented with nearly equivalent initial conditions. The growing rate of separation of the two orbits is clearly observable. In between the limiting cases of a purely regular and chaotic dynamics there are some systems exhibiting mixed behaviour [46].

2.2 Quantum and microwave billiards

In quantum mechanics the billiard dynamics is described by the stationary Schrödinger equation

$$(\Delta + k^2)\psi(x,y) = 0 \qquad k = \frac{p}{\hbar} = \frac{\sqrt{2mE}}{\hbar} \qquad (2.1)$$

with the boundary condition

$$\psi|_{\partial\mathcal{G}} = 0. \qquad (2.2)$$

Here, m and E are the mass and the energy of the particle, respectively, $\psi(x,y)$ is the wave function and $\partial\mathcal{G}$ is the boundary of the billiard domain $\mathcal{G}$ [47]. The potential is zero inside the billiard and infinite at the walls, which are impenetrable barriers. Set of discrete eigenvalues k_n and the corresponding eigenfunctions ψ_n constitute the solution of Eq. (2.1).

Let us consider a flat, cylindrical microwave cavity of height d with ideal conducting walls. The electromagnetic waves inside the billiard are described by the vectorial Helmholtz equation [48]. It reads

$$(\Delta + k^2)\vec{E}(\vec{r}) = 0 \quad \text{and} \quad (\Delta + k^2)\vec{B}(\vec{r}) = 0 \qquad (2.3)$$

---

[1] The Lyapunov exponent is defined as the natural logarithm of the average divergence rate of nearby points along the orbit.

for the electric and magnetic fields $\vec{E}$ and $\vec{B}$ inside the resonator, with $k = 2\pi f/c$ being the wave number, f the excitation frequency of the electromagnetic waves and c the velocity of light. In addition, the boundary conditions

$$\hat{n} \times \vec{E}(\vec{r})|_{\partial \mathcal{G}} = 0 \qquad \hat{n} \cdot \vec{B}(\vec{r})|_{\partial \mathcal{G}} = 0. \tag{2.4}$$

hold on the walls of the cavity. The vector $\hat{n}$ denotes the normal component to the surfaces of the resonator. We assume the z-axis to be perpendicular to the top and bottom surface. In our cylindrical geometry, the solutions of Eqs. (2.3) are categorized according to their polarization direction as transverse magnetic (TM) modes with $B_z=0$ or transverse electric (TE) modes with $E_z=0$. The electric field for a TM mode can be expressed as

$$E_z(\vec{r}) = E(x,y) \cos\left(\frac{n\pi z}{d}\right) \quad \text{with} \quad n = 0, 1, 2, \ldots \tag{2.5}$$

Substituting this expression into Eq. (2.3), the equation for the electrical field can be rewritten as

$$\left[\Delta + k^2 - \left(\frac{n\pi}{d}\right)^2\right] E_z(x,y) = 0 \tag{2.6}$$

with the boundary condition $E_z|_{\partial \mathcal{G}} = 0$.

Below the cutoff wave number $k_{\max} = 2\pi f_{\max}/c$ where $f_{\max} = c/2d$ is the corresponding cutoff frequency, the Eq. (2.6) has solutions only for $n = 0$ (TM$_0$-modes). Thus, there is no variation in z-direction of the electrical field and correspondingly a two dimensional description becomes possible.

Hence, the scalar Helmholtz equation for a quasi two dimensional cavity reads

$$(\Delta + k^2) E_z = 0 \tag{2.7}$$

with the boundary condition $E_z|_{\partial \mathcal{G}} = 0$ (cf. Eqs. (2.1) and (2.2)). For frequencies below f_c the scalar Helmholtz equation for the electrical field and the Schrödinger equation of a quantum particle inside a box with infinite hard walls are equivalent [47]. The cavities used in this work are 5 mm high corresponding to a cutoff frequency of 30 GHz. Due to the boundary conditions, two dimensional stationary waves can exist inside the cavity only for certain frequencies. As outlined in Sec. 1, such frequencies are called resonances and correspond to the cavity modes.

In conclusion, a microwave billiard provide a system to experimentally investigate properties of quantum billiards of the same shape since the time independent Schrödinger equation (2.1) and the scalar Helmholtz equation (2.7) for TM_0 modes are identical below the frequency f_{max}. Resonance frequencies and squared electrical fields in microwave billiards are fully equivalent to eigenvalues and probability densities in quantum billiards, respectively. Hence the investigation of wave phenomena in two dimensional electromagnetic cavities enables an experimental access to the study of quantum chaos.

3 Escape probabilities from classical billiards

One of the long standing problems of classical dynamics concerns the escape behaviour of an initial set of N particles from open classical billiards [49, 50]. Apart from the seminal work of Bauer and Bertsch [49], this issue was tackled e.g. in [51–53]. In [49] systems with chaotic dynamics are shown to decay exponentially, i.e. $\frac{dN}{dt} \propto e^{-\beta t}$ with the decay constant being proportional to the hole size Δ (see Fig. 2.1), whereas systems with regular dynamics are found to decay according to a power law, i.e. $\frac{dN}{dt} \propto t^{-\gamma}$.

Ray tracing simulations for the decay of two open classical billiards (circle and tilted stadium, see Fig. 2.1) have been performed. For this, a point-like particle is injected into the billiards at the opening position with $2.5 \cdot 10^6$ equidistant initial angles. It moves with the velocity of light. As soon as the particle escapes through the hole, a count is recorded with the time of flight it has taken to leave the system. This yields the time distribution of the escape probability. The sizes and shapes of the open billiards used for the numerical simulations coincide with those of the quantum billiards used in the experiments described in Sec. 5.2.2 (see Fig. 5.3 below). The classical and the quantum decay will be compared in Secs. 5.3.2 and 5.3.3. In the case of the circle, the opening sizes equal $10°$ and $20°$. In the case of the tilted stadium the opening is cut in the tilted side wall with lengths as large as the opening sizes in the circle billiard.

The classical escape probabilities thus obtained are shown in Fig. 3.1. The escape probability of the chaotic billiard shown in the upper panel displays an

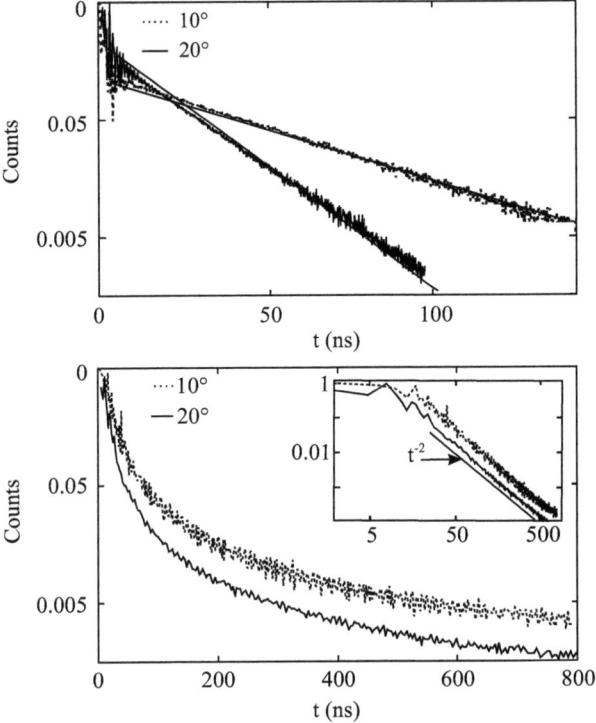

Fig. 3.1: Numerical simulations of the escape probabilities from the classical billiards shown in Fig. 2.1 (top: stadium, bottom: circle). In the upper panel the decay of the tilted stadium is plotted together with analytical predictions e^{-t/τ_W}. The lower panel shows the decay of the circular billiards. The inset demonstrates the observed t^{-2} decay behaviour.

oscillating structure with decreasing amplitude for times up to 10 ns which can be qualitatively understood as the time it takes for the ensemble of particles to "feel" the shape of the billiard [19]. For later times an exponential decay is observed in agreement with the results of [49]. The straight lines in the upper pannel correspond to the theoretical predictions with a decay constant equal to the inverse of the classical dwell time τ_W defined as

$$\tau_W = \frac{\pi A}{\Delta c}, \qquad (3.1)$$

where A is the area of the billiard. For an opening size of $10°$, $\tau_W = 23.68$ ns whereas for $20°$, $\tau_W = 11.85$ ns. A fit to an exponential function $e^{-\beta t}$ gives $1/\beta = 24.69 \pm 0.06$ ns for $10°$ and $1/\beta = 12.66 \pm 0.03$ ns for $20°$. In the case of regular dynamics, the escape probability follows a power-law decay $t^{-\gamma}$. A fit to an algebraic decay law yields $\gamma = 2.02 \pm 0.02$ for $10°$ and $\gamma = 1.97 \pm 0.04$ for $20°$. The double logarithmic plot in the inset of Fig. 3.1 shows the decaying curve t^{-2}. The decay curves exhibit a small sensitivity to the hole size. In [49], the decay exponent γ was also found to be equal to 2 and independent of the opening size.

4 Experiments with microwave billiards

4.1 Measurements at room temperature

The experiments described in this doctoral thesis are based on the electromagnetic excitation of a microwave cavity (also called microwave billiard). Figure 4.1 shows the parts of a microwave billiard. It is composed of three plates made of a conductor. Typically copper is used because of its small conductive losses and affordable cost. The middle plate defines the contour of the billiard while the upper and lower plate act as closings. The three plates are screwed together. In addition, wires of solder are put into grooves along the contour of the billiard in order to improve the electrical contact between the plates. A microwave cavity exhibits resonant behaviour, that is, the field inside of it oscillates at some frequencies, called its resonance frequencies, with much larger amplitude than at others.

The microwave power is coupled in and out through antennas which penetrate partially into the cavity. The antennas consist of a thin wire of about 0.5 mm diameter which is soldered in a holder. This is screwed on the upper plate as indicated in Fig. 4.1. The electromagnetic signal is generated by a vectorial network analyzer model Agilent PNA-L N5230A (in the following shorted as

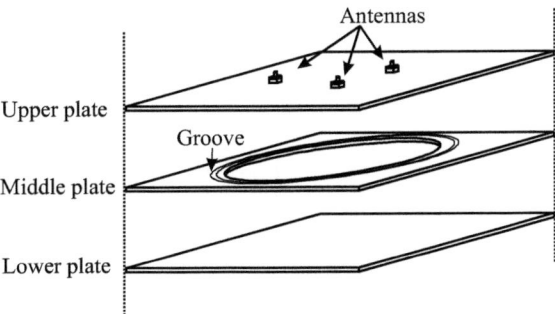

Fig. 4.1: Sketch of the modular assembly of a microwave billiard. The plates are screwed together. The hole in the middle plate defines the shape of the billiard. A thin groove for the solder wire is milled into the middle plate.

VNA). The electromagnetic power is led to one antenna wire (port a) via a coaxial conductor and fed into the cavity. In transmission measurements, the output signal is coupled out through another antenna (port b) and in reflection measurements via the same (port a). The ratio between the output power P_b and the input power P_a as well as the phase difference between both signals are recorded. This leads to the definition of the squared modulus of the scattering parameter as

$$|S_{ba}(f)|^2 = \frac{P_{out,b}(f)}{P_{in,a}(f)}, \qquad (4.1)$$

where P_n is the power at port n. A graphical representation of $|S_{ba}(f)|^2$ versus the excitation frequency is referred to as a frequency spectrum. Figure 4.2 shows a frequency spectrum of the closed circular cavity used in the experiments described in this doctoral thesis. It is measured at superconducting conditions. Resonances are clearly visible as peaks in the frequency spectrum. In the measurements described in this doctoral thesis the excitation frequency is varied from 0.05 to 25 GHz.

RF coaxial cables are used to transmit the microwave signal from the VNA to the antennas. The attenuation of the cables increases with the excitation frequency but generally does not exceed 5 dB/m. The cables are terminated with low-loss connectors (SMA 3.5 mm standard). Usually, a calibration of the whole

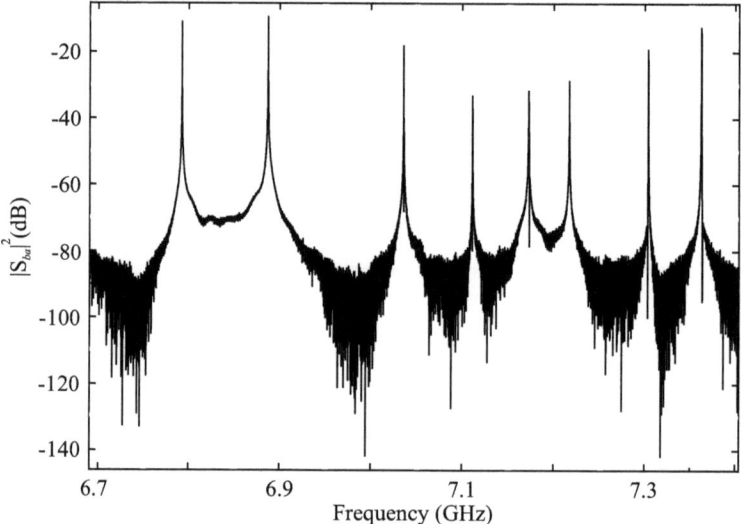

Fig. 4.2: Experimental transmission spectrum of a closed circular cavity at superconducting conditions. The plotted quantity is the absolute value of the the transmission amplitude of the scattering matrix. The maxima correspond to the resonances. These can be described in good approximation by a Breit-Wigner function.

setup is performed with an electronic calibration module in order to get rid of any influence of the measurement device and microwave cables.

Since antennas couple microwave power into and out of the cavity, they act as point-like perturbations [54]. The less deep the antennas penetrate into the cavity, the less they influence the field distribution. A clear disadvantage is that then less power is coupled into the system, and thus the resonance amplitudes diminish and they eventually cannot be distinguished from the noise level. In measurements at room temperature a compromise is achieved by choosing the penetration length of the antenna wires to be half of the cavity height, i.e. 2.5 mm [55]. As will be explained in Sec. 4.2, in the measurements at superconducting conditions in the framework of this doctoral thesis, the length is set to 0.5 mm.

The resonance frequencies, i.e. the positions of the resonances, can only be resolved if the resonance widths are considerably smaller than the separation between them (resonance spacings). An appropiate parameter to characterize the goodness of a resonance is the quality factor. This quantity measures the average number of oscillations the microwaves undergo before dissipating their stored energy in the cavity [48]. For a single resonance the quality factor turns out to be the ratio between the resonance frequency and the resonance width. For cavities excited at room temperature like those described in Sec. 6, the quality factors of the resonances lie between 10^3 and 10^4. According to the so called Weyl formula [56], the density of resonances increases with the excitation frequency. As a consequence, the resonances are resolvable only at low frequencies and thus the resonances parameters are determinable. At high frequencies, the resonances overlap and cannot be resolved anymore.

Close to the resonance frequency f_n of the nth isolated resonance, the transmission amplitude between two antennas located at points r_a and r_b in the cavity can be expressed in good approximation [57, 58] as

$$S_{ba} \sim \delta_{ab} - i \frac{\sqrt{\Gamma_a \Gamma_b}}{f - f_0 + i\frac{\Gamma_0}{2}}. \tag{4.2}$$

In this equation $\Gamma_{n,a}$ and $\Gamma_{n,b}$ are the partial widths related to the antennas a and b, and f_n and Γ_n denote the position and width (FWHM) of the resonance, respectively. Note that close to a resonance frequency, S_{ba} is essentially given by the usual Green function up to a certain frequency dependent factor [14]. This connection can be exploited in experiments in a number of different manners. Since $\Gamma_{n,a}$ is proportional to the square of the eigenfunction $\psi_n^2(r_a)$ of the closed billiard, the transmission probability between the two antennas is proportional to $\psi_n^*(r_a)\psi_n(r_b)$ at the nth cavity resonance. By fixing the position of one antenna, and varying the position of the second one inside the cavity, it is possible to extract the wave function by measuring S_{ba}. If the receiving antenna is moved to the outside of an open cavity the analogy between the microwave cavity and a quantum billiard is lost and the Green function does not correspond anymore to Eq. (4.2) but nevertheless the transmission amplitude remains proportional to $E_z(r_a)E_z(r_b)$. This will be illustrated by measurements shown in Sec. 6.

4.2 Measurements at superconducting conditions

For measurements at superconducting conditions all parts of a microwave cavity are coated with a thin layer (a few μm) of lead. Figure 4.3 illustrates the experimental setup used for these measurements. One or two microwave billiards are placed into a copper box which is brought into a cryostat. The box is precooled with liquid nitrogen to a temperature of 77 K. Then the nitrogen is extracted and the setup is cooled down to a temperature of 4.2 K by means of a thermal bath of liquid helium. Once the billiards reach the superconducting state at a critical temperature of 7.2 K, the resonance spectra are measured with a pressure of about 10^{-2} mbar in the box. The temperature can be kept stable for approximately 100 hours. Ideally, the resistance inside the walls of the cavity drops to zero abruptly when it is cooled down below 7.2 K. In reality, dissipation occurs due to thermal excitation of Cooper pairs by the RF field and also due to a small residual resistance [59]. The antennas attached to the resonator are connected to the cryostat lid through semi-rigid RF cables. These are led inside vacuum roughing hoses. Vacuum feedthroughs on the cryostat lid provide the joint with the cables which are connected to the VNA. A software programm controls the operation of the VNA. A microwave switch connected to the VNA ports enable the automatic gathering of data sets from different antenna combinations. The helium level, temperature and pressure in the copper box and in the cryostat are continuosly controlled during the measurements.

Figure 4.4 shows a part of two transmission spectra taken between two antennas placed in a circular billiard with an opening on the boundary (see inset of Fig. 4.4). The two curves correspond to measurements performed at 4.2 K (solid) and 77 K (dashed). The decrease of the resonance widths in the spectrum measured at 42 K is significant (in particular for the right peak) yielding a great increase of the quality factor. As a consequence complete sequences of resonance data (position and width) may be obtained for closed resonators up to high frequencies. For the right peak of Fig. 4.4 at 77 K, $Q \approx 1500$ while at 4.2 K, $Q \approx 84000$. The characteristic width of the resonances measured under superconducting conditions is mainly due to the electrical coupling of the antenna wires to the cavity. Resonances taken in cavities at normalconducting conditions additionally acquire a width due to dissipation of microwaves in the copper plates.

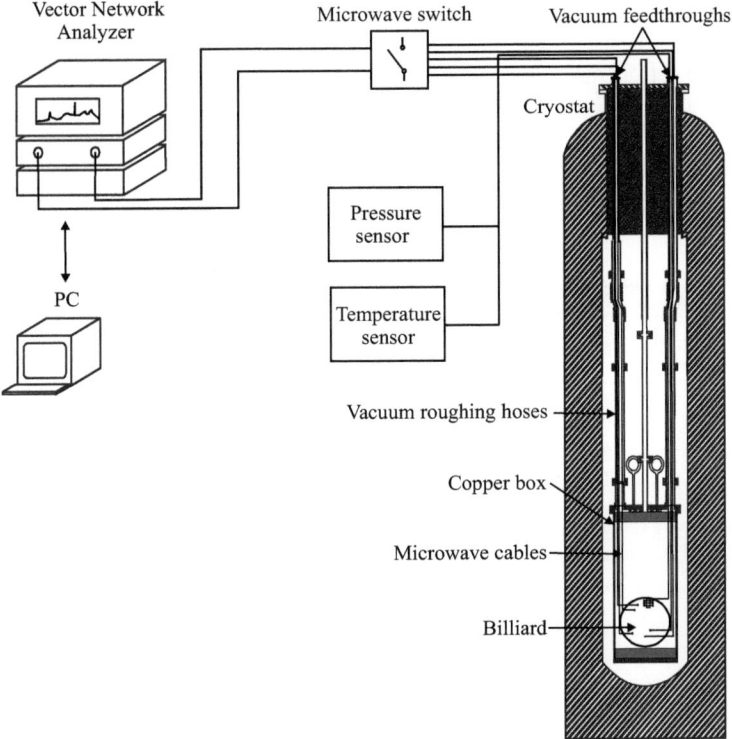

Fig. 4.3: Schematic picture of the experimental setup for measurements at superconducting conditions (from [60]). The billiard is brought into an evacuated copper box and the whole device inserted into a cryostat filled with liquid helium. The billiard is connected to the VNA through coaxial cables and a microwave switch. The measured data are transferred to a PC during the measurement.

We moreover observe a shifting of the resonance frequencies between both spectra which is the result of the thermal contraction of the cavity (see Fig. 4.4).

In the next chapter measurements with open microwave billiards at superconducting conditions are presented.

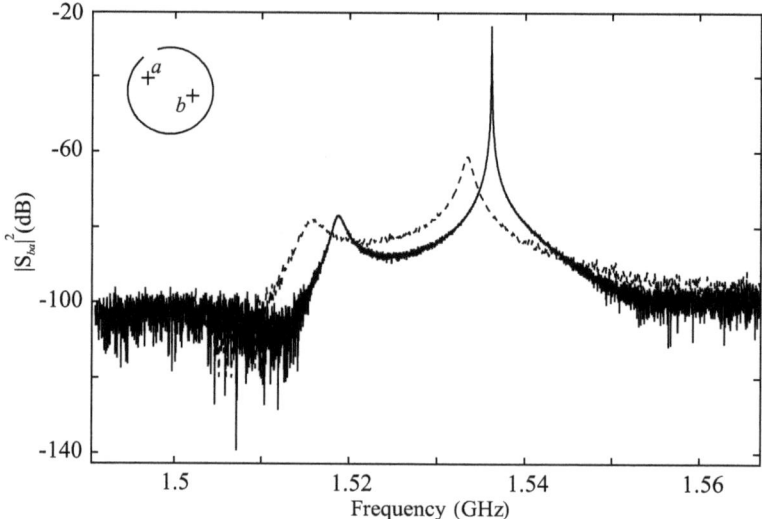

Fig. 4.4: Transmission spectra taken between two antennas placed in a circular billiard with an opening on the boundary (see inset). The crosses a and b indicate the positions of the antennas. The solid curve corresponds to a measurement at superconducting conditions whereas the dashed one to a measurement at normal conducting conditions.

5 Survival probability in open quantum billiards

5.1 Description of open quantum systems

5.1.1 Survival and decay probabilities

In reality, a quantum system is not isolated but interacting with its environment and has to be considered as an open system. This interaction leads to a mixing of the states, usually called decoherence, and to an energy exchange between system and environment which is typically known as dissipation.

As explained in Sec. 2.2 an exact mapping between the measured quantities and the eigenstates of the Schrödinger equation (2.1) is possible in microwave experiments allowing to view the microwave resonator as an "analog computer" solving the Schrödinger equation. Since these experiments measure the outcome of scattering events it seems reasonable to adopt the formalism derived in the theory of quantum mechanical scattering. A very useful object for analyzing the process of quantum scattering is the scattering matrix that relates the outgoing wave amplitudes to the incoming ones [61]. In the more general description of the Hamiltonian approach (initially derived by Weidenmüller and Mahaux in the framework of nuclear physics [62]) the cavity is coupled to the outside by leads and the internal motion manifests in the properties of the scattering matrix [58, 63].

Assuming M open channels in the leads and N discrete states for the bound system, the $M \times M$ scattering matrix can be written in the form [62]

$$S_{cc'}(E) = \delta_{cc'} - 2\pi i \left\langle c, E | W^\dagger (E - H + i\pi W W^\dagger)^{-1} W | c', E \right\rangle \tag{5.1}$$

with

$$\mathcal{H} = H - i\pi W W^\dagger. \tag{5.2}$$

Here H stands for the N-dimensional self-adjoint Hamiltonian describing the bound system, the states $|c, f\rangle$ refer to the continuum states, f is the frequency of the incoming waves, and W denotes an $M \times N$ operator that contains matrix elements W_{ic} coupling the bound mode i to the open channel c. No continuum-continuum coupling is permitted in the model.

The solution of the whole scattering system depicted in Fig. 5.1 is written as $\Phi = \begin{pmatrix} \psi \\ \phi \end{pmatrix}$, where ψ is the internal wave function represented by the N-component vector $(\psi_1, ..., \psi_N)^T$ and ϕ is a M-component vector $(\phi_1, ..., \phi_M)^T$, with

$$\phi_n \propto (A_n e^{ik_\parallel^{(n)} x} + B_n e^{-ik_\parallel^{(n)} x}) \sin(k_\perp^{(n)} y). \tag{5.3}$$

the solution of the Schrödinger equation inside the lead of width d for the open channel n. The coordinate x is defined along the center line of the waveguide with origin at the attached point to the cavity and the coordinate y is perpendicular to it. In Eq. (5.3) $k_\parallel^{(n)} = \sqrt{k^2 - \left(k_\perp^{(n)}\right)^2}$ and $k_\perp^{(n)} = \frac{n\pi}{d}$ denote the two components of the wave vector k, one transversal and one longitudinal to the lead boundaries.

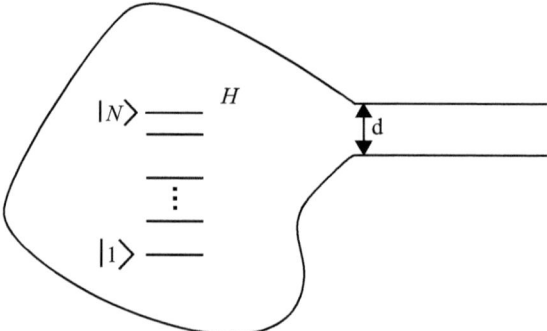

Fig. 5.1: Sketch of a scattering system (from [64]). A shaped cavity containing N bounded states and described by a Hamiltonian H is attached to an infinite rectangular lead of width d.

The amplitudes A_n and B_n are the nth components of the vectors $\mathbf{A}$ and $\mathbf{B}$ of incoming and outgoing wave amplitudes, respectively. As mentioned previously, they are linked by the S-matrix, i.e. $\mathbf{B} = S\mathbf{A}$. The index n defines the number of nodes of the wavefunction in transversal direction to the lead boundary or equivalently, the number of quantum mechanically open channels and is equal to the largest integer less or equal to kd/π.

The probability $P(t)$ that the system remains bounded at time t is defined as

$$P(t) = \text{Tr}(\rho(t))/\text{Tr}(\rho(0)) \tag{5.4}$$

with $\rho(t)$ being the density matrix of the system at time t. In the following $P(t)$ is referred to as the survival probability. For a closed system $P(t) = 1$. The time evolution of the density matrix is given by the effective Hamiltonian $\mathcal{H}$ according to

$$\rho(t) = e^{i\mathcal{H}t}\rho(0)e^{-i\mathcal{H}^{\dagger}t}. \tag{5.5}$$

An explicit expression for $P(t)$ in terms of the scattering matrix is given as

$$P(t) \propto \int\int df_1 df_2 \frac{e^{i(f_2-f_1)t}}{f_2-f_1} \sum_{n,m} A_n^* A_m \left[(S^{\dagger}(f_1)S(f_2))_{nm} - \delta_{nm}\right], \tag{5.6}$$

for a complete derivation see [64] or [65].

According to the continuity equation, the total current leaking out of the cavity shown in Fig. 5.1 is connected to the survival probability by

$$j(t) = -\frac{\partial P(t)}{\partial t}. \tag{5.7}$$

Then from Eqs. (5.6) and (5.7) follows

$$j(t) \propto \int\int df_1 df_2 e^{i(f_2-f_1)t} \sum_{n,m} \left[A_n^* A_m (S^\dagger(f_1)S(f_2))_{nm} - \delta_{nm} \right], \tag{5.8}$$

which shows that the decay of an open quantum system is essentially determined by the Fourier transform of the autocorrelation function $S^\dagger(f_1)S(f_2)$. This result was initially derived by Harney, Dittes and Müller [?]. If the wave packet impinges the system through only one of the open channels (as actually is the case in the experimental realization), say the channel a, then $A_a = 1, A_{b \neq a} = 0$ and the current reduces to

$$j(t) \propto \int\int df d\epsilon e^{i\epsilon t} \sum_a \left[(S^\dagger(f)S(f+\epsilon))_{aa} - \delta_{aa} \right], \tag{5.9}$$

where a change of variables $(f_1, f_2) \to (f, \epsilon = f_2 - f_1)$ has been carried out. Since in the experiments no reflection elements of the scattering matrix (i.e. S_{aa}) are measured, the product of (5.9) is restricted to the transmission parameters $S_{ab}, b \neq a$, where a and b symbolize the positions of the antennas. Then the expression used for the current leaking out of the cavity reads

$$j(t) \propto \int\int d\epsilon dE e^{i\epsilon t} \left[\sum_{a \neq b} S_{ba}(f) S_{ba}^*(f+\epsilon) \right] \tag{5.10}$$

where we have assumed $S_{ab} = S_{ba}$. Furthermore, since the expression (5.10) reduces to the Fourier transform of a correlator, the so called Wiener-Khinchin theorem can be applied [66]. This theorem states in its general form

$$\mathcal{F}\left[\int_{-\infty}^{\infty} g^*(x')g(x+x')dx' \right] = |\mathcal{F}[g]|^2. \tag{5.11}$$

where $\mathcal{F}[g]$ denotes the Fourier transform of the function g. The computation of the integrals of Eq. (5.10) in terms of the absolute squared value of the fast Fourier transform of the measured complex S-matrix elements is considerably faster than evaluating the twofold integral. This is one of the reasons why we decided to compute the leaking current given in Eq. (5.10) instead of the survival probability given in Eq. (5.6). Summarizing, in the following the current leaking out of the cavity is simply determined through

$$j(t) \propto \sum_{a \neq b} |\mathcal{FT}[S_{ba}(f)]|^2 j(t) = \sum_{a \neq b} |\tilde{S}_{ba}(t)|^2, \qquad (5.12)$$

where the sum extends over several antennas combinations.

5.1.2 Resonance widths

The poles of the scattering matrix S are associated with the formation of resonant states. From Eq. (5.1) of Sec. 5.1.1 it is clear that the formation of resonances is closely related to the dynamics of the system which is governed by H. Correspondingly, the poles of S represent quasi-bound states to which bound states of a closed system are converted due to the coupling to the continuum. The eigenvalues of $\mathcal{H}$ can be written as $\mathcal{E}_n = f_n - i\Gamma_n/2$ where f_n and Γ_n are the frequencies and the widths of the resonances, respectively. As shown in Sec. 4.1, the line shape of every resonance state is well described by a Breit-Wigner function. Accordingly, the contribution from each mode decays exponentially in time (deviations are expected for very short and long times [67, 68]). Close to the nth isolated resonance,

$$\mathcal{F}[S_{ba}(f)] \propto \sqrt{\Gamma_{na}\Gamma_{nb}} e^{-\pi\Gamma_n t}. \qquad (5.13)$$

Thus the resonance widths Γ_n play the key role in the decay since they are inversely proportional to the lifetimes of the corresponding resonance states. The distribution of partial widths is also closely related to the decay behaviour.

If the resonances are well isolated, the decay is given by the average of the (appropriately weighted) decay laws of the N individual resonances,

$$P(t) = \frac{1}{N} \sum_{n=1}^{N} w_n e^{-\pi\Gamma_n t}, \qquad (5.14)$$

where the weight w_n is related to the probability of excitation of the resonance n at time $t=0$. As found in [69], for long times the sum will be eventually dominated by the slowest-decaying modes which, according to Eq. (5.13), decay exponentially. Furthermore, the contribution of each resonance to $P(t)$ can be substituted by an average over all possible widths Γ, the weight function being given by the corresponding width distribution $\rho(\Gamma)$, yielding

$$P(t) = \bar{\Gamma}^{-1} \int_0^\infty \Gamma \rho(\Gamma) e^{-\pi \Gamma t} d\Gamma. \quad (5.15)$$

where $\bar{\Gamma}$ denotes the average width. In the region of well isolated resonances, the widths of a generic quantum scattering system with classically chaotic dynamics are distributed according to the Porter-Thomas distribution [14]

$$\rho_{PT}(\Gamma) = (2\pi \Gamma \bar{\Gamma})^{-1/2} e^{-\Gamma/2\bar{\Gamma}} \quad (5.16)$$

which leads to a survival probability

$$P(t) = \bar{\Gamma}^{-1} \int_0^\infty \Gamma \rho_{PT}(\Gamma) e^{-\pi \Gamma t} d\Gamma = (1 + 2t\bar{\Gamma})^{-3/2} \quad (5.17)$$

for one open channel. This result has been generalized to a few open channels and sufficiently large times by Harney, Dittes and Müller. Using the analytic expression based on RMT for the S-matrix autocorrelation function $(S^\dagger(f_1) S(f_2))_{nm}$ (see Eq. (5.6)) in the framework of compound nucleus reaction theory (initially derived by Verbaarschot, Weidenmüller and Zirnbauer [70]), they showed that in the regime of isolated resonances [?]

$$P(t) \propto (1 + 2t\bar{\Gamma})^{-M-1/2}, \quad (5.18)$$

with M the number of open channels. Ranging from isolated resonances to strongly overlapping ones, a lot of work has been devoted to explore the widths distribution behaviour in quantum mechanical systems with chaotic classical dynamics [64, 71–73]. In these references a crossover from the Porter-Thomas distribution for the regime of isolated resonances to a broad power-like distribution typical for the regime of overlapping resonances was found. Experimentally, the distribution of partial widths was investigated with a superconducting microwave cavity with the shape of a stadium billiard [16, 74] and inside chaotic partially open systems [75]. In [16] the measurements were performed at superconducting conditions with a resonator with chaotic dynamics whose boundary is closed, such that the microwave power can exit the cavity only via the antennas and the widths were found to follow a Porter-Thomas distribution. In the systems studied in the present work, however, the resonators have an opening in the boundary, such that the individual eigenvalues are not resolvable since the vast majority of the measured resonances are overlapping. This complicates critically the check of the statements posted in [64, 71–73]. The largest widths are characteristic for those states that leave the system after a short time (the ones this work is interested in) but at the same time they overlap and thus interfere strongly with the neighbouring resonances. In conclusion, the decay properties of open quantum systems can be discussed in terms of statistical properties of the scattering matrix related to the effective Hamiltonian introduced in Sec. 5.1.1.

5.2 Experiment

5.2.1 Design of the resonators

In order to experimentally determine the decay probability in microwave cavities with regular and chaotic dynamics, different billiards with openings on the boundary have been used. To investigate the regular dynamics a circular billiard of 120 mm radius was fabricated from copper. The (closed) circle billiard is a paradigm for regular systems. The eigenfunctions of the quantum circle billiard are given as combinations of Bessel functions and trigonometric functions. Two quantum numbers, n_r and m account for the number of nodes of the wavefunctions in radial and azimuthal directions, respectively. Accordingly, they are called radial and azimuthal quantum numbers. The size of the opening in the boundary of the circular cavity ranges from an angle α of 5° to 20° in intervals of 5° (see Fig. 5.2). At the opening the system is coupled to a triangular lead which permits the quantum states to decay in (see Fig. 5.2). Absorption material for microwaves is glued to the attached lead walls in order to avoid microwave reflections back into the cavity. The absorber material used consists of urethan foam sheets EPP-51 impregnated with carbon (from the company ARC Technologies). For this system measurements of intensity distribution have been performed (see Sec. 5.2.2).

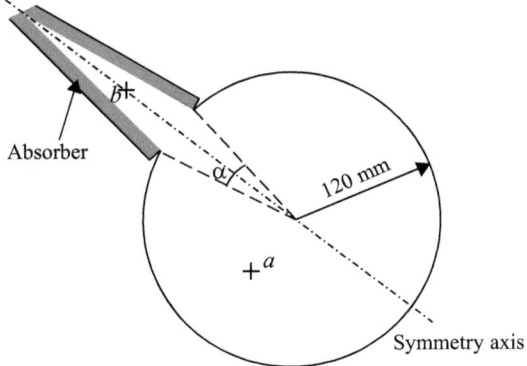

Fig. 5.2: Sketch of the circular billiard with attached lead. The axis transversal to the opening (dash-dotted line) defines the symmetry line. The antenna positions are indicated with crosses. In the attached lead walls absorption material has been placed in order to avoid reflections back into the cavity.

Due to the proximity of the antenna b to the absorption material glued to the walls (see Fig. 5.2), only a very small current leaking out of the cavity was measured with the antenna b. Therefore a new enlarged lead was attached to the circular billiard for the measurements of the resonance spectra. The left panel of Fig. 5.3 illustrates the dimensions of the new cavity. Again the absorption material of the lead is attached to the walls to prevent reflection of power back into the cavity. The area and perimeter of the cavity are 452.4 cm^2 and 75.4 cm, respectively, whereas the area of the lead is about 132 cm^2. A desymmetrized tilted stadium was also designed and manufactured from copper to study the decay of a chaotic system (see right panel of Fig. 5.3). The opening is cut in the tilted side wall and its size is chosen as large as the opening sizes $d = R\alpha$ in the circle billiard. The area of the stadium billiard and the attached lead is the same as for the circular cavity in order to obtain comparable resonance densities. The perimeter of the cavity is equal to 88.7 cm.

Both systems were subsequently plated with a layer of lead of about 100 μm thickness for measurements at superconducting conditions.

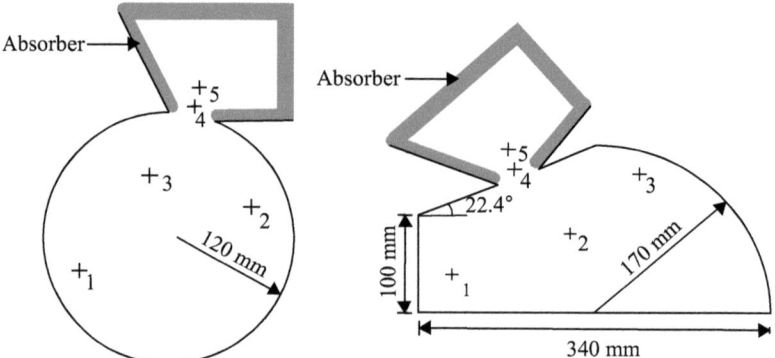

Fig. 5.3: Scheme of the cavities with the attached leads. The area of the cavities and leads as well as the opening sizes are the same for both systems. The antenna positions are indicated with crosses. Absorption material is attached to the lead walls.

5.2.2 Measurements of intensity distributions of the open circle billiard

In order to visualize the current leaking out of the open circular cavity depicted in Fig. 5.2, measurements of the electric field intensity were performed with the so called perturbation body method using a cylindrical perturber made of magnetic rubber (for a detailed explanation see page 17 of Ref. [76]). The introduction of the perturbation body into the cavity changes the boundary conditions and correspondingly the resonance frequencies are sligthly shifted [54]. According to the Maier-Slater theorem [77], the frequency shift is proportional to the squared field at the position of the perturbation. The perturbation body is guided by an external magnet across the billiard and lead areas with a spatial resolution of 2 mm. The measurements were performed inside a thermostat box at about 300 K in order to avoid thermal fluctuations which can alter the resonance frequencies due to thermal expansion of the cavity. Thus by measuring the shift of the resonance frequencies we determine the electric field intensity distribution, which corresponds directly to the squared modulus of the quantum mechanical wave function (see Sec. 2.2).

Due to the opening in the boundary the rotational symmetry of the circle is broken and the degenerate quantum states (in the circle states with $m \neq 0$) are split into doublets. As Fig. 5.4 shows, the corresponding eigenfunctions are symmetric or antisymmetric with respect to the symmetry axis defined by the opening position [78]. For the symmetric states the intensity in the lead is appreciable. On the contrary no intensity is measured outside of the circular cavity for the long-lived states. Accordingly one partner of the doublet is strongly affected by the opening (symmetric state) and displaced to lower frequencies while the other one is barely modified (antisymmetric state). This fact is clearly reflected in the widths of both states. The width of the symmetric state increases considerably since it is strongly coupled to the lead and thus decays fast. Such symmetric states are thus called short-lived. The frequency splitting of doublet partners increases with the opening size and also with frequency.

The influence of the opening is especially strong for modes with high field intensity at the opening. The so called whispering gallery modes, i.e. states with

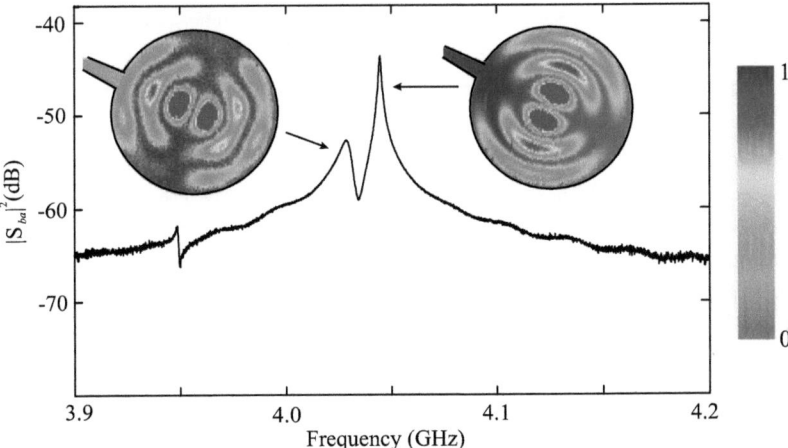

Fig. 5.4: Part of the frequency spectrum corresponding to the split states of the doublet with quantum numbers $n_r = 3$ and $m = 1$. The opening angle is 20°. Measured electric field intensity distributions are also plotted next to the resonances.

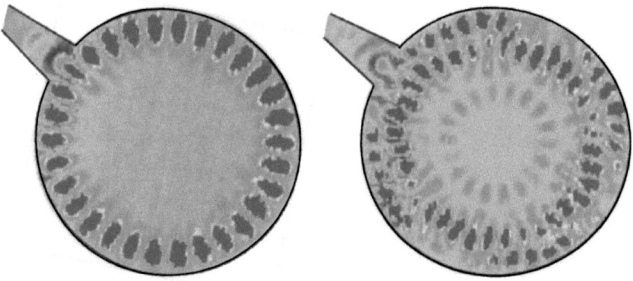

Fig. 5.5: Two examples of intensity distributions of whispering gallery modes in the open circular billiard with opening angle 20°. The resonance frequencies are 7.921 GHz (left) and 10.943 (right). For the color scale see Fig. 5.4.

small n_r but large m, as well as single states with $m = 0$ but large n_r are intensely coupled to the lead [79, 80]. In [80] such modes were found to be short-lived states coupling strongly to an attached lead and supporting direct transport processes. In Fig. 5.5 two experimental intensity distributions of whispering gallery modes are plotted. The state represented in the left panel has a resonance frequency of 7.921 GHz and its radial quantum number n_r is 1. Close to the opening the patches are deformed since there the field intensity partly leaves the system. The state represented in the right panel has a resonance frequency of 10.943 GHz and n_r is 2. The rotational symmetry of the whispering gallery mode is broken and the field distribution adopts a polygonal structure. This is widely observed in other wave functions whose intensity distributions are located close to the boundary.

The microwave power exiting the cavity into the lead can be detected by measuring the transmission between one antenna placed inside the resonator and one antenna inside the lead. Figure 5.6 shows frequency spectra from 0.05 to 30 GHz for each one of the investigated opening sizes where the emitting antenna a is put inside the system and the receiving one b in the lead. For each opening there is clearly no transmission below a certain frequency. This can be understood by using instead of the attached triangular lead of Fig. 5.2 a lead with parallel metalic walls with Dirichlet boundary condition (2.2). Namely, with the notation introduced in Sec. 5.1.1, in a lead with parallel walls the wave vector k can be decomposed into two components, one transversal and one longitudinal to the parallel walls. Then the longitudinal component of the wave vector inside the lead is given as

$$k_{\parallel}^{(n)} = \sqrt{k^2 - (\frac{n\pi}{R\alpha})^2}. \qquad (5.19)$$

Only waves with a discrete and finite number of nodes transversal to the lead boundary can travel through the lead. These correspond to a quantized traverse wave vector $k_{\perp}^{(n)} = n\pi/R\alpha$. Thus for $n = 1$ there exists a cutoff frequency f_{cut}

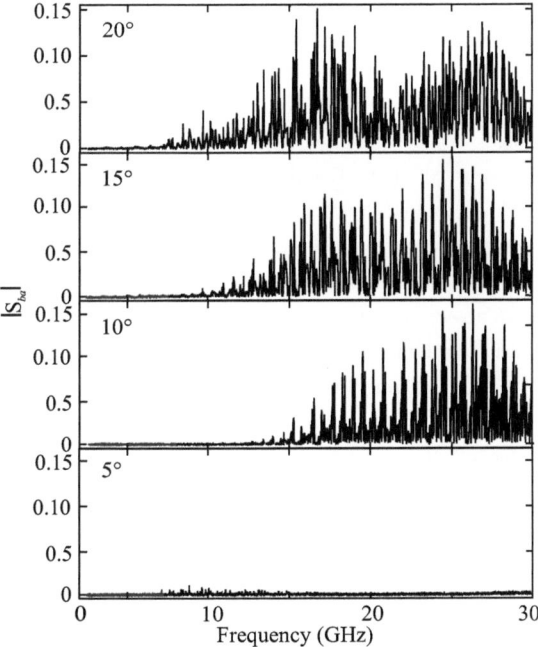

Fig. 5.6: Measured frequency spectra for each of the considered opening sizes. The emitting antenna a is put inside the system and the receiving one b in the lead. Below the cutoff frequency defined by each of the openings no power is transmited between the antennas (see the main text).

below which there are no open channels. The cutoff frequency is given by

$$f_{\text{cut}} = \frac{c}{2R\alpha}. \tag{5.20}$$

The respective cutoff frequencies f_{cut} for the opening sizes 20°, 15°, 10° and 5° are 3.75, 7.5, 15 and 30 GHz, respectively. They are in very good agreement with the threshold frequency for transmission observed in Fig. 5.6. These results confirm that despite coupling the circular cavity to a triangular lead, this can be effectively treated as a lead with parallel walls. Thus in the case of 5° opening, we detect only some exponential decaying evanescent modes.

5.2.3 Measurements of frequency spectra

The decay behaviour, especially for chaotic systems, has attracted much attention in the last years. It has been intensively studied in the framework of chaotic scattering both theoretically [22, 81] and experimentally in microwave resonators [82, 83]. The results presented in [82] are concerned with the decay behaviour of a microwave cavity with chaotic dynamics attached to a single open waveguide. The survival probability was found to decay exponentially. One essential aspect in [82] is the large absorption of microwave power in the cavity walls at room temperature, which may be modelled by a large number of ficticious additional channels [84–87]. Thus the contribution of the absorptive losses to the decay of the resonances has to be accounted for since it disguises the pure decay through the opening of the billiard. This has not been considered in [82]. Indeed, an accurate determination of the survival probability demands measurements at superconducting conditions to minimize the absorptive losses (see Sec. 4.2) which has not been performed so far.

In this section the results of the measurements of frequency spectra at superconducting conditions are presented. The aim was the determination of the temporal decay behaviour of quantum systems with regular and chaotic dynamics in order to establish their common features as well as their differences Experimentally, transmission spectra between two antennas are measured between 0.05 and 25 GHz with a frequency step of 15 kHz. This frequency step size suffices to resolve the narrowest resonances. Five antennas are put in every system (see

Fig. 5.3). Three of them are located inside the cavities and two inside the leads. For both the circle and the tilted stadium two different opening sizes (equivalent to 10° and 20° in the circular system) have been used. About 1000 states lie in the measured frequency range.

Figure 5.7 compares the time spectrum of open cavities measured at room temperature and at 4.2 K. The antennas are placed inside the cavities. The absolute value of the transmission parameter between both antennas is plotted (obtained by means of a Fourier transformation of the whole spectrum, see Sec. 6.4) which describes the temporal decay behaviour of the microwave power inside the system. An offset time corresponding to the time the electromagnetic signal needs for travelling through the cables has been removed and $\tilde{S}_{ba}(t)$ has been normalized such that $|\tilde{S}_{ba}(0)| = 1$. At 300 K there is no remarkable difference between the decays of both systems. We conclude that the decay at room temperature is effectively governed by absorption in the cavity walls. For the superconducting cavities, the decay is considerably slower (especially for the regular cavity) yielding a richer decay structure with several time scales.

Tle lowe panel of Fig. 5.8 shows the frequency spectrum of two of the low-lying resonance states ($m = 3$, $n_r = 1$) for the open circular cavity. One solid curve

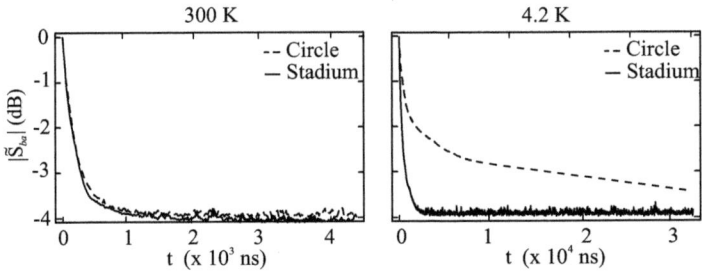

Fig. 5.7: Comparison of time spectra from the measured open cavities taken at room temperature (left panel) and at superconducting conditions (right panel). The absolute value of the Fourier transform of S_{ba} is plotted versus time, which gives the temporal behaviour of the microwave power inside the system. Note the different time scales of left and right panel.

corresponds to an opening size of 20° and the other one to 10°. For comparison we also show in the upper panel the resonance spectrum measured with the closed circular cavity as solid line. The frequency spectra were measured with two antennas located inside the cavity (see inset). The change of the boundary condition due to the opening of the billiards splits the degeneracies and shifts the resonance positions to lower frequencies with respect to the resonance frequency of the closed system. A situation like that presented in Fig. 5.4 is achieved. The right partner of the doublets has a much smaller width than the left one, i.e. the hole in the boundary leads to a lift of the degeneracies in the widths and frequencies of the resonances. In contrast, in a closed circular cavity, the breaking of the rotational symmetry leads only to a separation of the degenerate resonances frequencies. The short-lived state is shifted to lower frequencies with increasing hole size. The widths and resonance frequencies of the presented doublets are summarized in Tab. 5.1. While the widths of the antisymmetric states are about one order of magnitude larger than the corresponding one in the closed system, the widths of the symmetric ones are three orders of magnitude larger. The small widths of long-lived states explain the long-time tails in the circular system (cf. Fig. 5.7).

As expected, the width of the left partner of the considered doublet increases with the size of the hole since it mainly contributes to the decay probability through the opening. However the opposite behaviour is observed for the right partner. This strange finding is also found for many other low-lying doublets and could be related to the mechanism of resonance trapping [88–90]. In this contraintuitive effect the resonance widths of all states should first increase with increasing coupling strength to the channels but at a critical coupling they decrease again for some of the states. An appropiate scenario to observe such a phenomenon may be two closely lying resonance states coupled to one common decay channel [88] like the present case of degenerate quantum states split into doublets. The effect of resonance trapping has been experimentally observed in [91].

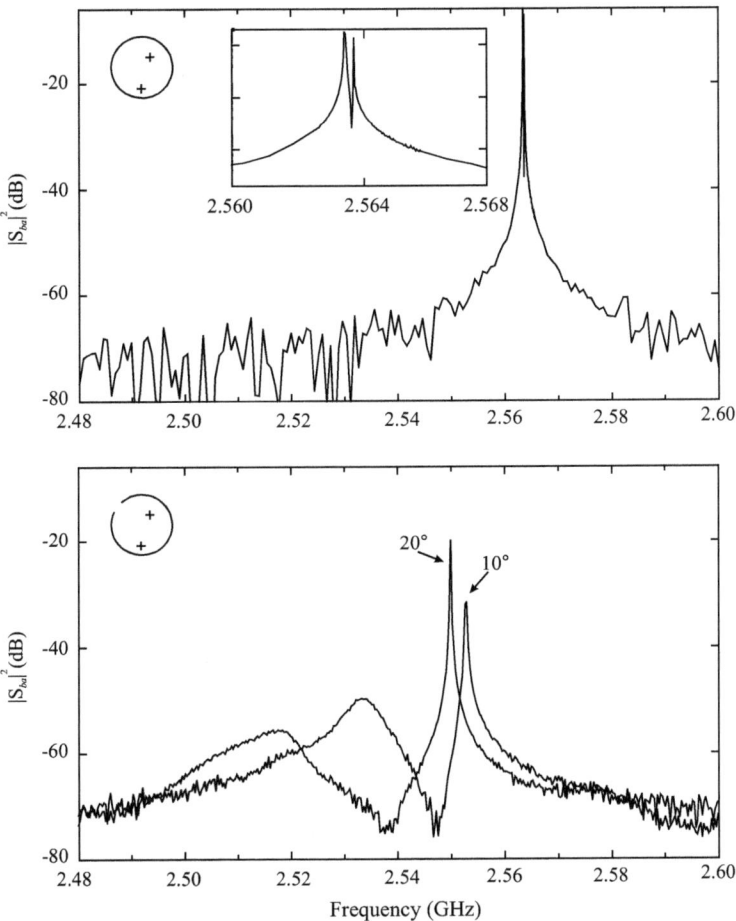

Fig. 5.8: A resonance doublet of the closed circular billiard (upper panel). The inset shows a magnification of the resonance doublet. The same resonance doublet of the open circular billiard with an opening size of 10° and 20° (lower panel). The antennas lie inside the circular billiard in both cases (see insets).

Tab. 5.1: Resonance frequencies and widths of the doublets represented in Fig. 5.8.

	f_μ(GHz)	Γ_μ (kHz)
Short-lived state 20°	2.5162	11.4368
Short-lived state 10°	2.5334	6.7459
Long-lived state 20°	2.5503	0.0761
Long-lived state 10°	2.5521	0.1543
State 1 closed circle	2.5634	0.0090
State 2 closed circle	2.5637	0.0085

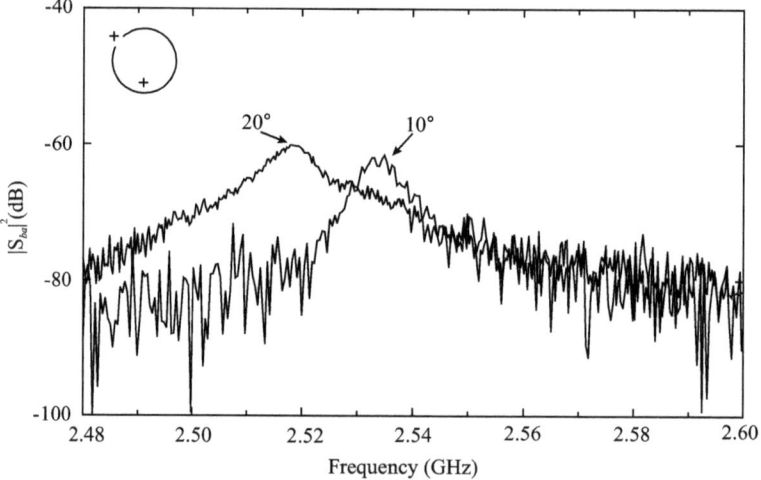

Fig. 5.9: The same resonance doublet as in Fig. 5.8 with an opening of 10° and 20°. The receiving antenna is located outside the circular billiard (see inset).

The same doublet as in Fig. 5.8 is plotted in Fig. 5.9, but this time the receiving antenna lies outside the cavity (see inset). The amplitude of the right partners is drastically reduced and mixes with the noise level, i.e. the antisymmetric states do not leave the system since they correspond to modes very weakly coupled to the antenna placed outside (compare with Fig. 5.4). Thus to investigate the decay of modes which leave the system quickly through the opening, such antenna combinations are used (see Secs. 5.3.2 and 5.3.3). The isolated resonance corresponding to the second lowest-lying state in the open stadium billiard (10° and 20° opening size) is shown in Fig. 5.10. The antennas are located inside the system (see inset). A large opening manifests itself in wider resonances. This behaviour is observed for all states. For certain modes, however, the increase of widths is anomalously large.

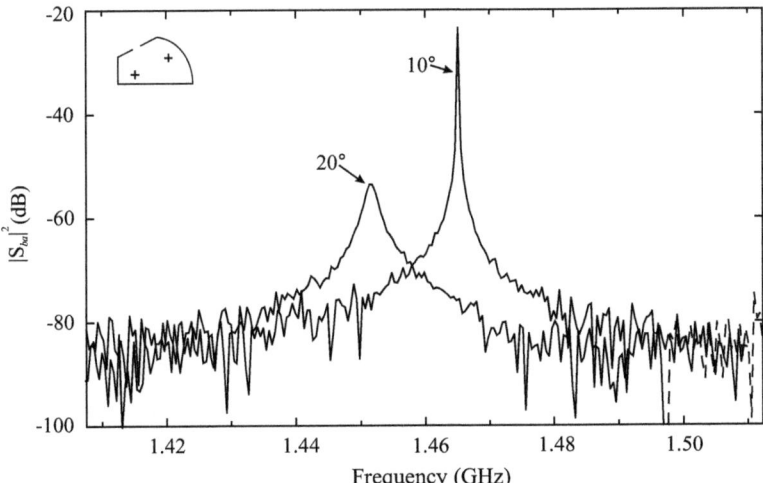

Fig. 5.10: Second lowest-lying state of the tilted stadium with an opening of 10° and 20°.

5.3 Analysis and interpretation

5.3.1 Temporal decay behaviour of single resonances

Figure 5.11 shows the decay behaviour for six single resonances (dots). The ones shown in the left panels are obtained from measurements with the open circle billiard whereas the ones in the right panels result from measurements with the open stadium billiard. The opening size is the same for both cavities, i.e. 10° in the circular billiard.

According to Eq. (5.13), an exponential decay is expected for an isolated resonance. However, some states display deviations in the long-time region. For instance, for the resonances shown in the lowest panels of Fig. 5.11 we observe three regimes with different decay behaviour: For short times it is exponential in very good agreement with Eq. (5.13). Then, a subsequent algebraic decay is observed up to very long times ($\approx$ 20000 ns). Finally the signal mixes with the noise level. The origin of the algebraic decay behaviour can be understood as follows: the Fourier transform of the Breit-Wigner function which describes a single resonance with frequency f_0 and width Γ_0 (see Eq. (4.2)) is given by

$$\tilde{S}_{ba}(t) = \int_{f_{\min}}^{f_{\max}} \frac{\sqrt{\Gamma_a \Gamma_b}}{f - f_0 + i\frac{\Gamma_0}{2}} e^{-2\pi i f t} df \qquad (5.21)$$

with $f_{\min}$ and $f_{\max}$ being the lower and the upper limit for the Fourier transform. For short times we have

$$|\tilde{S}_{ba}(t)| \sim \frac{\sqrt{\Gamma_a \Gamma_b}}{2\pi} e^{-\pi \Gamma_0 t}. \qquad (5.22)$$

The long-time behaviour is determined by performing partial integration of Eq. (5.21),

$$\tilde{S}_{ba}(t) = -\frac{\sqrt{\Gamma_a \Gamma_b}}{2\pi i t} \frac{e^{-2\pi i f t}}{f - f_0 + i\frac{\Gamma_0}{2}} \Big|_{f_{\min}}^{f_{\max}} - \int_{f_{\min}}^{f_{\max}} \frac{\sqrt{\Gamma_a \Gamma_b}}{2\pi i t} \frac{e^{-2\pi i f t}}{(f - f_0 + i\frac{\Gamma_0}{2})^2} df$$

$$= -\frac{\sqrt{\Gamma_a \Gamma_b}}{2\pi i t} \frac{e^{-2\pi i f t}}{f - f_0 + i\frac{\Gamma_0}{2}} \Big|_{f_{\min}}^{f_{\max}} + \mathcal{O}(t^{-2}). \qquad (5.23)$$

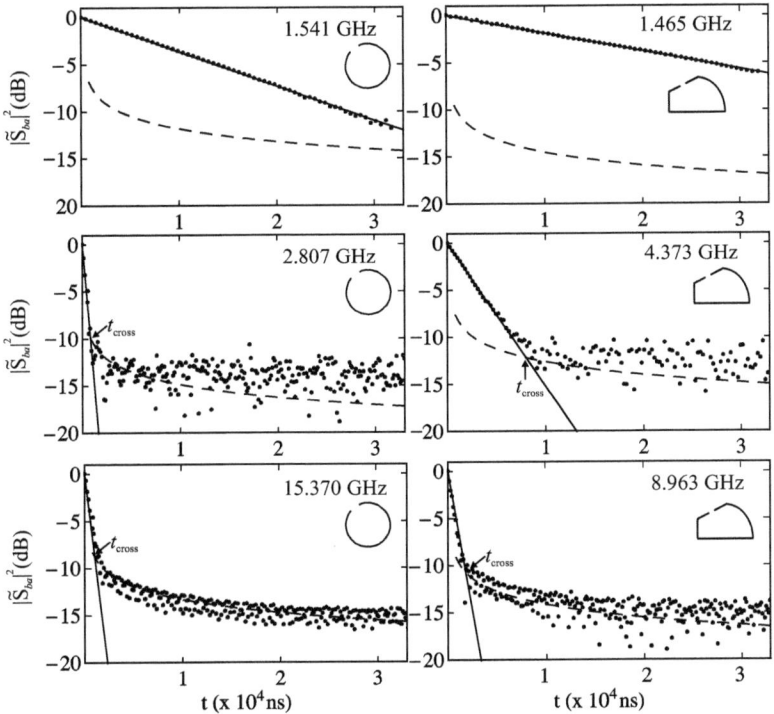

Fig. 5.11: Decay of isolated resonances in spectra measured for the two billiards (left column: circle; right column: stadium) with an opening of 10° (dots). The solid line corresponds to $e^{-2\pi\Gamma_0 t}$ and the dashed line to $A(f_0, f_{\min}, f_{\max}, \Gamma_0)/t^2$, see Eq. (5.24) in the main text.

For very long times we can neglect the term $\mathcal{O}(t^{-2})$. Taking the logarithm of the squared modulus of $\tilde{S}_{ba}(t)$, we obtain

$$\ln\left(|\tilde{S}_{ba}(t)|^2\right) = \ln(\frac{A(f_0, f_{\min}, f_{\max}, \Gamma_0)}{t^2}) - 2\pi\Gamma_0 t \qquad (5.24)$$

where $A(f_0, f_{\min}, f_{\max}, \Gamma_0) = \frac{\Gamma_a \Gamma_b}{4\pi^2}\left(\frac{1}{(f_{\min}-f_0)^2+(\Gamma_0/2)^2} + \frac{1}{(f_{\max}-f_0)^2+(\Gamma_0/2)^2}\right)$. Accordingly, the solution of the trascendental equation

$$\frac{A(f_0, f_{\min}, f_{\max}, \Gamma_0)}{t^2} = e^{-2\pi\Gamma_0 t} \qquad (5.25)$$

defines a time scale where the decay behaviour changes from exponential to algebraic. This time scale is called t_{cross} and is indicated in the panels of Fig. 5.11. The two terms of the right hand side of Eq. (5.24), i.e. the exponential and algebraic decay curves, are plotted in the panels of Fig. 5.11 as solid and dashed lines, respectively. Apart from the upper panels, t_{cross} is clearly observed. It should be noted that the t^{-2} decay law is not related to the dynamics of the system but only to the values of $f_{\min}, f_{\max}$ and Γ_0. In the upper panels of Fig. 5.11, the resonances decay too slowly to observe the t^{-2} decay due to the small widths Γ_0 (0.03 MHz for the open circular billiard and 0.06 MHz for the open stadium billiard). Correspondingly, t_{cross} lies at times not accessible in the experiments. For the plots in the middle part of Fig. 5.11, t_{cross} lies at 750 ns (circle) and 8300 ns (stadium). The algebraic decay describes the data well up to about 15000 ns for the circle and to 5000 ns for the stadium. For longer times the noise level is reached.

5.3.2 Temporal decay behaviour of the open circle billiard

According to Eq. (5.12), the decay probability is obtained as an average over the time spectra for antenna combinations. For each antenna combination, only a finite number of modes is excited. Since the mode excitation, i.e. the partial width and the amplitude of excitation, and thus the resulting decay function depend on the antenna position, the sum over all antenna combinations behaves as an ensemble average.

In the measured systems there are three antenna combinations with antennas inside the billiards (1-2, 1-3, 2-3, see Fig. 5.3), and six antenna combinations with one antenna inside and the other outside the cavity (1-4, 1-5, 2-4, 2-5, 3-4 and 3-5, see Fig. 5.3). In the following, the combinations are denoted as I-I (inside-inside) and I-O (inside-outside), respectively. As seen in Sec. 5.2.3, the I-O combinations mainly measure the decay of those modes which leave the cavity. In contrast, the I-I combinations predominantly detect modes which remain trapped for long times inside the billiards. Figure 5.12 shows the decay probabilities in three different frequency intervals for I-I combinations (left panels) and I-O ones (right panels) in the open circle billiard. The opening sizes are 10° (dots) and 20° (circles). Three different frequency ranges for the Fourier transform of Eq. (5.12) are analyzed: 2-5 GHz (top), 12-15 GHz (middle) and 20-23 GHz (bottom).

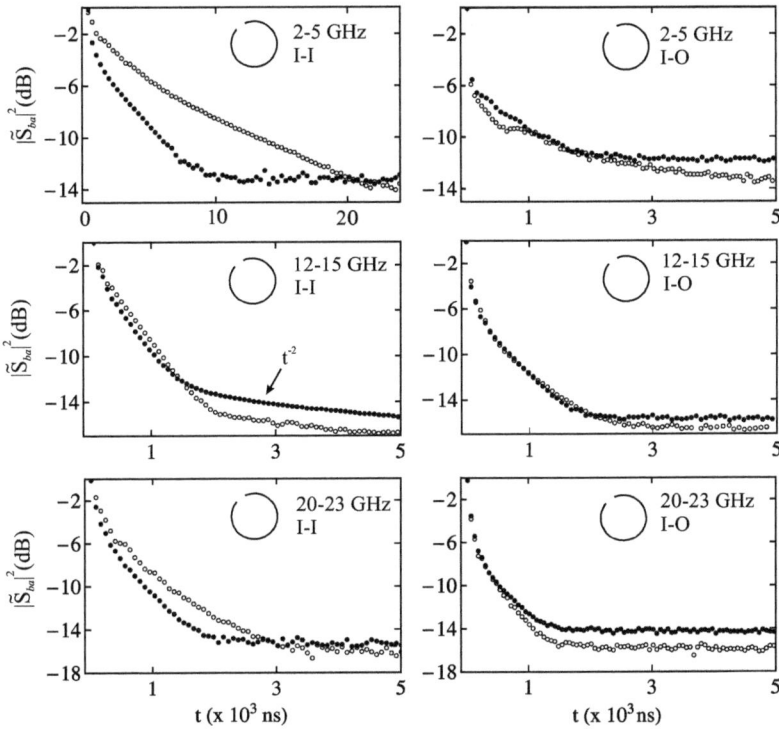

Fig. 5.12: Decay probabilities in three different frequency intervals for I-I combinations (left) and I-O (right) ones in the open circle billiard. The opening sizes are 10° (dots) and 20° (circles). Note the different time scale for the top left panel.

For the frequency interval corresponding to 2-5 GHz and I-I combinations (left top panel), the decay for the billiard with 20° opening is considerably slower than for the one with 10° opening. This is certainly surprising but it can be understood as follows: In the low-lying region of the spectra we observe states split into doublets (see Figs. 5.4 and 5.8). As Tab. 5.1 shows, the width of the antisymmetric state (i.e. trapped state in the cavity) of the doublets is smaller for the 20° opening than for the 10° opening. This is related to the mechanism of resonance trapping and explains the anomalous decay behaviour since the trapped states mainly contribute to the decay probability for I-I combinations, having the largest resonance amplitudes (see Fig. 5.8 and Refs. [88–90]). Furthermore, for I-O combinations, the antisymmetric state of the doublets cannot be excited (see Fig. 5.9), i.e. the symmetric states predominantly contribute to the decay. Since the widths of such states are larger for the 20° opening than for the 10° opening, the decay is faster for the former in the interval 2-5 GHz (at least up to 1000 ns, see right top panel). In the frequency ranges 12-15 GHz and 20-23 GHz, the decay for combinations I-I with an opening of 10° (i.e. decay of trapped modes) is slightly faster than the one corresponding to 20°. However, for I-O combinations we do not observe remarkable differences.

For the frequency range 12-15 GHz (left middle panel), a t^{-2} decay law like for an isolated resonance (see Sec. 5.3.1) is observed for the I-I combinations. This can be explained as follows. First, among the states which dominate the decay behaviour for long times (i.e. the ones with the smallest widths), there is one whose amplitude is much larger than that of the others. Second, the frequency of this dominant resonance lies close to 15 GHz (maximal frequency of the interval). Acordingly, the function $A(f_0, f_{\min}, f_{\max}, \Gamma_0)$ from Eq. (5.24) increases and the t^{-2} decay law of such a resonance is reinforced. We remark that if the maximal and minimal frequencies of the interval are shifted by 0.5 GHz, the t^{-2} decay behaviour vanishes.

5.3.2.1 Long-time behaviour in the open circular billiard

According to Eq. (5.14), we can express the decay probability as a sum of exponential functions with appropiate coefficients. This is confirmed with the results presented in this section. Figure 5.13 again shows the decay probabilities for the opening angle of 10° (I-I combinations) in the frequency windows 2-5 GHz, 12-15 GHz and 20-23 GHz for times up to the t^{-2} decay or the noise level. Beside the experimental results (dots), three additional curves are shown. In each panel the solid line results from the decay function obtained from the sum

$$P(t) = \sum_{\mu} A_{\mu} e^{-2\pi \Gamma_{\mu} t}, \qquad (5.26)$$

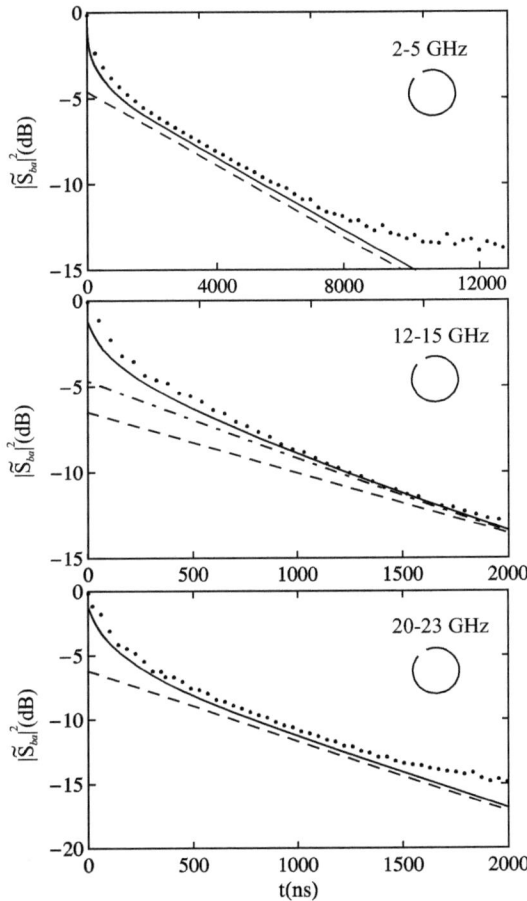

Fig. 5.13: Decay probabilities for I-I combinations and opening size 10° (dots). The sum (5.26) over exponentially decaying functions (solid line), the exponential function with the smallest decay constant, i.e. width (dashed line), and the sum over the four resonances with the smallest widths from Eq. (5.26) (dash-dotted line) are also represented. Note that the curves are shifted with respect to each other for the comparison of the long-time behaviour.

where the resonance parameters A_μ and Γ_μ are extracted from the measured resonance spectra in the corresponding frequency ranges. Only resonances were taken into account whose shape is well described by a Breit-Wigner function, i.e. those ones with small enough resonance widths. The dashed line is obtained from a single exponential function $e^{-\beta t}$ with the decay constant β equal to that corresponding to the smallest width Γ_μ in the sum of Eq. (5.26). Moreover, for the interval from 12 to 15 GHz (middle panel) the dash-dotted line represents the sum over the four resonances with the smallest widths in the sum of Eq. (5.26).

In the frequency window 2-5 GHz, the narrowest resonance determines the overall time behaviour for sufficiently long times (after 2000 ns). Correspondingly, the decay probability behaves exponentially in the long-time region. In the frequency window 12-15 GHz, however, the state with the smallest width does not describe the long-time behaviour. Instead, the sum over the four states with the smallest widths determines the decay behaviour in the interval 500-1800 ns. The decay is again exponential for sufficiently long times ($>$ 1200 ns) since the contribution of one resonance is dominant with respect to the others. Accordingly, beside the widths, the amplitudes of resonances definitely play a decissive role in the decay. In the frequency window 20-23 GHz the term with the smallest width again suffices to describe the experimental data as in the case of the top most panel.

In Fig. 5.14 frequency spectra for the intervals shown in Figs. 5.12 and 5.13 are depicted. They are taken with one I-I antenna combination. The lines indicate the position of the resonances associated with eigenstates with $m = 1, 2, 3$ of the corresponding closed circular billiard and the insets show the intensity distribution. Due to the contraction of the cavities at low temperatures (see Sec. 4.2), the exact radius of the circular cavity is not known. The eigenvalues are obtained as the zeroes of the Bessel functions $J_m(k_\mu R)$. Thus the exact value of the radius R of the circle billiard can be estimated from the determined zeroes, i.e. from the position of the low-lying resonances in the spectrum. As can be seen from the spectrum from 2 to 5 GHz (upper panel of Fig. 5.14), this leads to a good agreement between calculated eigenvalues and experimental resonance frequencies. However, for high frequencies (12-15, 20-23 GHz) the position of the resonances in the spectra are shifted to larger frequencies with respect to the calculated values.

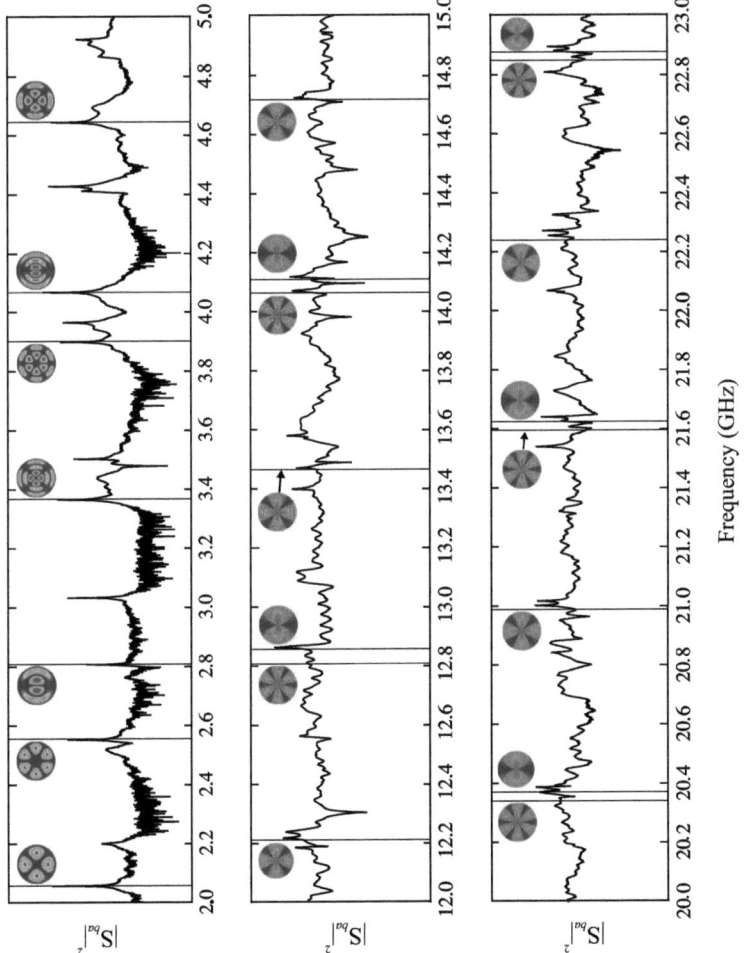

Fig. 5.14: Frequency spectra in the windows 2-5, 12-15 and 20-23 GHz taken with antennas inside the open (20°) circular billiard. The lines and the plots indicate the theoretical eigenvalues and the intensity distributions of the associated eigenstates of the corresponding closed circular billiard, respectively.

We attribute these discrepancies to the shift of the eigenvalue positions due to the opening. Indeed, as demonstrated in Fig. 5.8 the change of the boundary condition when removing part of the wall causes a frequency splitting of the nearly degenerate doublet partners which increases with the opening size and with frequency. Taking these discrepancies into account we observe that in the high-frequency regions (12-15 GHz and 20-23 GHz) most of the resonances with the smallest widths correspond to states with $m = 1, 2, 3$ and high n_r. They clearly stick out from the oscillating structure of strongly overlapping resonances and yield the dominant contribution to the slow decay of trapped modes inside the billiard. This reflects the observed prominent contribution of certain narrow resonances to the decay behaviour (see Fig. 5.13). It persists even after averaging over several antenna combinations. Correspondingly, the exponential decay law observed in certain (long) time regimes for the three curves presented in Fig. 5.12 is related to these trapped modes [88, 92].

5.3.2.2 Short-time behaviour in the open circular billiard

The decay behaviour for the measured open circle billiards is presented up to 200 ns ($\approx 150\, t_H{}^2$) in Fig. 5.15 . Time spectra obtained from the frequency ranges 2-5 GHz (dots), 12-15 GHz (circles) and 20-23 GHz (stars) measured with I-O antenna combinations are plotted as dashed line. The upper (lower) panel shows the result for 10° (20°) opening. Beside the experimental curves, the corresponding classical decay is depicted as solid line (see Sec. 3). For high-frequency intervals the quantum mechanical decay behaviour tends to the classical one in accordance with the correspondence principle [93] (especially for 20-23 GHz). Accordingly, the decay in the high frequency regions is well described by a t^{-2} law.

5.3.2.3 Length spectra

A connection between the spectra of quantum billiards and the periodic orbits of the corresponding classical system can be obtained by means of a semiclassical approach based on trace formulas [12]. The effect of classical POs on spectral

---

[2]The Heisenberg time t_H is defined as $\hbar/\langle d \rangle$ with $\langle d \rangle$ being the mean level spacing between resonances. In the present study it is given by $A f_{\max}/2\pi c^2$, where A denotes the area of the billiard and $f_{\max}$ the maximal frequency for the corresponding interval.

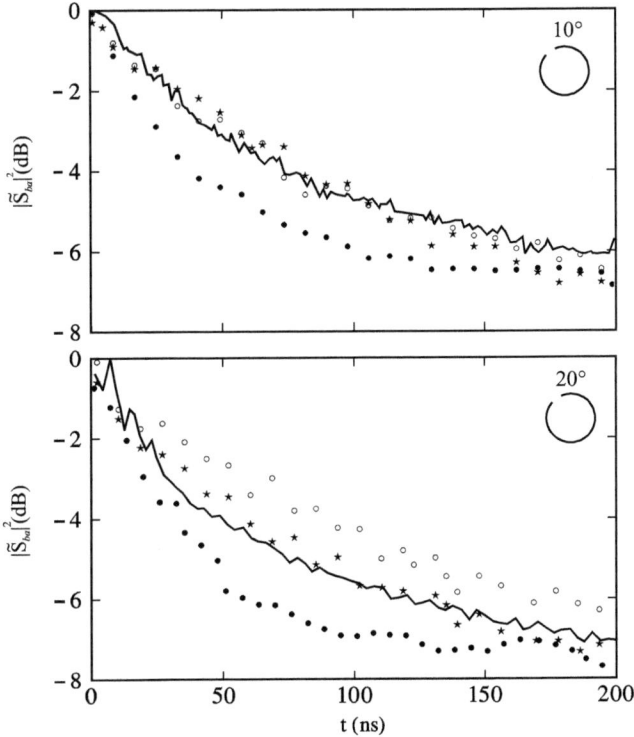

Fig. 5.15: Decay probabilities for the open circle billiard with an opening of 10° (top) and 20° (bottom) measured with I-O combinations for the intervals: 2-5 GHz (dots), 12-15 GHz (circles), 20-23 GHz (stars). The corresponding classical decay is also plotted as solid line.

properties of the quantum mechanical system is observable in the Fourier transform of the fluctuating part of the level density,

$$\hat{\rho}_{fluc}(l) = \int_{k_{\min}}^{k_{\max}} dk e^{ikl} [\rho(k) - \rho_{\text{Weyl}}(k)], \tag{5.27}$$

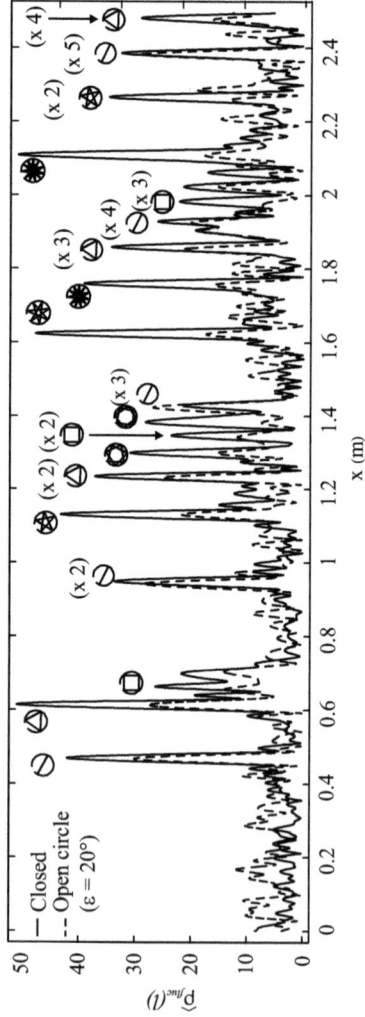

Fig. 5.16: Length spectrum of the closed circular cavity (solid line) and the open circular cavity with 20° opening (dashed line). The corresponding frequency interval is 2-18 GHz.

where $\rho(k)$ is the measured level density of the system, $\rho_{\text{Weyl}}(k) \propto k$ is its smooth part obtained from Weyl's formula [93], and $[k_{\min}, k_{\max}]$ is the wave number interval in which the data are taken. In classical circular billiards, periodic orbits (POs) are typically stars and polygons (see [94] for some examples of such orbits). In Fig. 5.16, experimental length spectra $\tilde{\rho}_{fluc}(l)$ of the closed and open (20°) circular billiards are presented. The positions of the peaks agree with the lengths of the POs and their amplitudes are related to the stability of such orbits. In a similar context, the time spectrum of the conductance of a microscopic circular junction with two leads shows peaks which are interpreted as the classical paths followed by electron waves [31]. The amplitudes of POs corresponding to polygons with a high number of vertices and to stars are manifestly supressed in the open cavity. This suggests that these orbits are eminently affected by the hole. In contrast, the amplitude of the diameter orbit is approximately unaffected in all repetitions. Experimental studies of relevant POs in a quantum annular billiard have been also carried out in [95] showing that orbits with marginal stability are robust to small perturbations (in this case the positions of the asymmetrically placed disk of the annular setup).

5.3.3 Temporal decay behaviour of the open stadium billiard

Figure 5.17 shows the temporal decay behaviour for three different frequency regions for I-I antenna combinations (left panel) and I-O ones (right panel) in the open stadium billiard. The opening sizes used are equivalent to the length corresponding to angles 10° (dots) and 20° (circles) in the circle billiard. As in Sec. 5.3.2, the frequency ranges for the Fourier transform of Eq. (5.12) are chosen as 2-5 GHz (top most panel), 12-15 GHz (middle panel) and 20-23 GHz (bottom panel). In contrast to the open circle billiards, the decay is slower for the billiard with an opening of 10° in all three frequency ranges as intuitively expected, especially for the frequency interval 2-5 GHz[3].

---

[3]In the interval 2-5 GHz the sharp distinction in the decay behaviour for the two opening sizes reflects the wide differences in the widths of resonances, see Fig. 5.10.

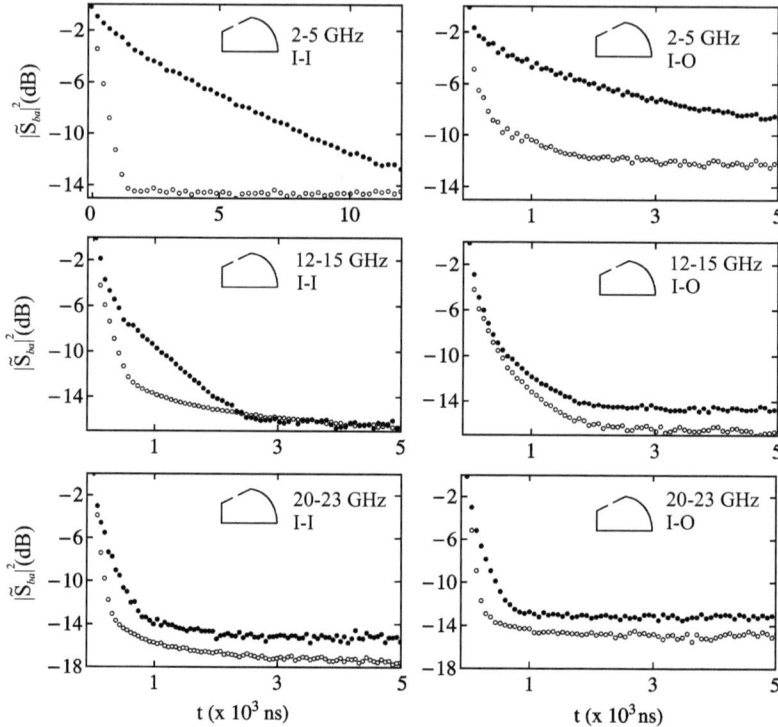

Fig. 5.17: Decay probabilities in three different frequency intervals for I-I (left) and I-O (right) antenna combinations in the open stadium billiard. The opening sizes are equivalent to 10° (dots) and 20° (circles) in the circle system.

5.3.3.1 Long-time behaviour in the open stadium billiard

The same analysis as in Sec. 5.3.2.1 for the circle billiard is performed for the open stadium billiard. Figure 5.18 shows again the temporal decay for the opening angle of 10° (I-I combinations) in the frequency windows 2-5, 12-15 and 20-23 GHz. The time intervals are chosen such that the long-time behaviour before reaching the noise level is observable.

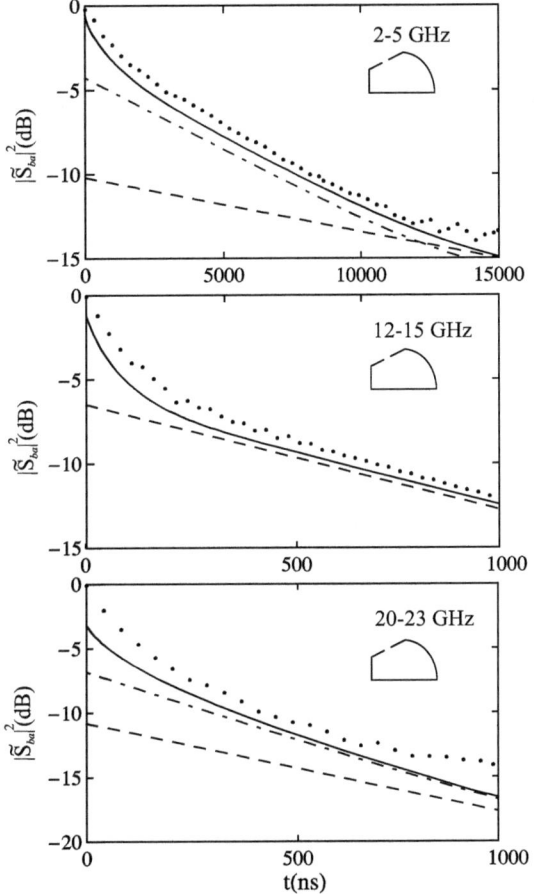

Fig. 5.18: Decay probabilities for I-I combinations and opening size 10° (dots). The sum (5.26) over exponentially decaying functions (solid line), the exponential function with the smallest decay constant, i.e. width (dashed line), and the sum over the four resonances with the smallest widths from Eq. (5.26) (dash-dotted line) are also represented. Note that the curves are shifted with respect to each other for the comparison of the long-time behaviour.

The solid lines in Fig. 5.18 represent the decay function obtained from the sum of Eq. (5.26). The dashed lines correspond to single exponential function $e^{-\beta t}$ with decay constant β equal to the smallest width in the sum of Eq. (5.26). For the case of the interval from 2 to 5 GHz (middle panel) and 20-23 GHz (lower panel), the dash-dotted curves additionally represent the sum over the four resonances with the smallest widths in the sum of Eq. (5.26). In the frequency window 2-5 GHz, the decay of the narrowest resonance (dashed line) does not describe the long-time behaviour. This is a similar situation as for the frequency interval 12-15 GHz in the open circle billiard shown in Fig. 5.13. The sum over the four resonances with the smallest widths approximates the decay function well for long times. In the present case, however, the four narrowest resonances have similar amplitudes and widths. Consequently, there is no dominant term in the sum of Eq. (5.26) such that the complete sum describes the experimental data better than one or a few summands. Hence, the decay deviates from an exponential behaviour. In the frequency window 12-15 GHz, however, the decay curve of the resonance with the smallest width suffices to describe the long-time behaviour. In the frequency window 20-23 GHz, the sum over the four resonances with the smallest widths gives again a better agreement with the data. The long-time decay differs clearly from an exponential decay.

Figure 5.19 shows the decay function for the stadium billiard with an opening of 20° and antenna combinations I-O. The frequency ranges for the Fourier transform are chosen such that the opening size allows one (3-6 GHz), two (6-9 GHz) and three (9-12 GHz) open channels. It should be remarked that in the present case there are no real channels in the sense of a rectangular waveguide (see Fig. 5.1). However, the openings define similar cutoff frequencies. According to the result derived by Harney, Dittes and Müller (Eq. (5.18)), the decay probability for a chaotic scattering system with isolated resonances, i.e. the time derivative of the survival probability, is given by

$$j(t) \propto (1 + 2t\bar{\Gamma})^{-M-3/2} \tag{5.28}$$

with M being the number of open channels, i.e. an algebraic decay-law is expected. A function of the form $A(1+Bt)^{-\gamma}$ with A, B and γ as fit parameters describes the experimental data shown in Fig. 5.19 well in certain time intervals, especially in the intermediate time regime where the prediction of Harney, Dittes and Müller,

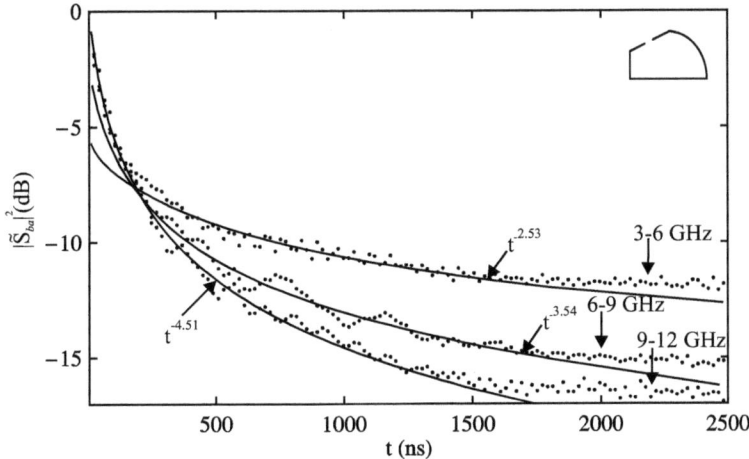

Fig. 5.19: Decay probabilities for the stadium billiard with an opening of 20° and one (3-6 GHz), two (6-9 GHz) and three (9-12 GHz) open channels (dots). The algebraic fittings describe quite well the decay curves from the experimental data (solid lines).

i.e. formula (5.18), is expected to be valid. Moreover the fitted decay exponents γ are in good agreement with the ones expected according to Eq. (5.28). For the frequency interval 3-6 GHz we obtain $\gamma = 2.53 \pm 0.05$ in the time interval 340-1540 ns, for 6-9 GHz, $\gamma = 3.54 \pm 0.07$ in the time interval 210-1540 ns and for 9-12 GHz, $\gamma = 4.51 \pm 0.11$ in the time interval 200-1540 ns. Hence, an algebraic decay is observed and the expected dependence on the number of open channels is obtained.

5.3.3.2 Short-time behaviour in the open stadium billiard

The decay behaviour for the measured open stadium billiards is presented in Fig. 5.20 up to 100 ns ($\approx 75\, t_H$). The frequency ranges for the Fourier transform are 2-5 GHz (dots), 12-15 GHz (circles) and 20-23 GHz (stars). Furthermore, the corresponding classical decays are depicted as solid line (see Sec. 3). Similarly to the behaviour in the open circle billiards, for the frequency window 20-23 GHz, the quantum mechanical decay behaviour tends to the classical one. As exposed in Sec. 3, the classical decay is described by an exponential function $e^{-\beta t}$ with β the inverse of the classical dwell time, see Eq. (3.1). In [82] this decay behaviour is also found for an elbow billiard with one open channel. However, as pointed out in Sec. 5.2.3, the large absorption in the cavity walls in the experiments of [82] raises the question whether the considered scattering system corresponds to a one-channel case.

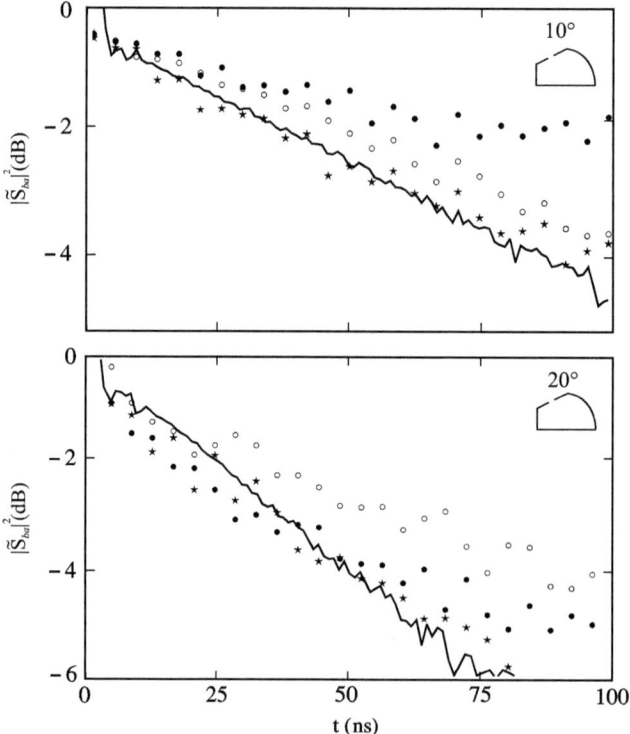

Fig. 5.20: Decay probabilities for the open stadium billiard with an opening of 10° (top) and 20° (bottom) measured with I-O combinations for the intervals: 2-5 GHz (dots), 12-15 GHz (circles), 20-23 GHz (stars). The corresponding classical decay is also plotted as solid line.

5.4 Conclusions

The quantum-mechanical decay probability is computed from the resonance spectra by means of the formula (5.18) suggested by Harney, Dittes and Müller [?]. It corresponds to an approximation of the analytic expression for the S-matrix autocorrelation function derived by Verbaarschot, Weidenmüller and Zirnbauer in the context of compound nucleus reaction theory [70].

For the shortest times accessed experimentally, the decays tend to the corresponding classical analoga in the semiclassical limit for both the open circle and the open stadium billiard (power-law $t^{-\gamma}$ for the former (see Fig. 5.15) versus exponentially fast $e^{-\beta t}$ for the latter (see Fig. 5.20) where β is intimately related to the parameter which specifies the classical dynamics, i.e. the classical dwell time τ_W). In the present experiments the decay exponent γ is obtained as 2 in agreement with the classical behaviour described in [49] and in experiments with dielectric cavities [23]. Similarly, in atom-optics billiards a power-law escape behaviour was observed for atoms trapped in an open circular billiard up to about 20 τ_W [25]. In the same reference a nearly exponential behaviour is obtained for the stadium billiard up to 10 τ_W. In the experiments described in the present section we observe the same behaviour as in [25] up to similar time scales for both, the circle and the stadium. However, for longer times quantum mechanics smears out the clear difference between chaotic and regular motion in striking contrast to the classical decay. In [34] the decay rate for long times (above 100 t_H) is shown to obey a clear power law that depends on the number of open channels.

In the long-time region and for I-I antenna combinations, we find for both the chaotic (stadium billiard) and the regular (circle billiard) situations frequency intervals where an exponential long-time tail is observed due to the dominance of some long-lived modes (see Fig. 5.13). These have smalls widths and large excitation amplitudes. For the circular billiard these modes correspond to sharp resonances with large n_r but small m[4] (see Fig. 5.14). For the stadium billiard, however, a weighted sum over a few resonances describes the experimental data better in some frequency intervals (see Fig. 5.18). Thus, there the decay behaviour deviates from a pure exponential one. In other frequency intervals the current

---
[4]As explained in Sec. 5.2.2 n_r and m are the radial and azhimutal quantum numbers, respectively.

probability eventually transitions to an exponential decay in accordance with results obtained in [69].

For the stadium billiard with I-O antenna combinations we observe that in the long-time region the decay probability shows an essentially nonexponential behaviour despite the fact that single resonances decay exponentially (see Fig. 5.17). In the regime of isolated resonances, the decay probability is well described by a power-law behaviour (see Eq. (5.28)) obtained by replacing the sum in Eq. (5.26) by a weighted average over the resonance widths. The expected dependence on the number of equivalent open channels proposed by Harney, Dittes and Müller in the frame of quantum chaotic scattering is achieved for one, two and three open channels (see Fig. 5.19). In contrast, for some time regimes in the circle billiard the decay probability exhibits exponential features (see Fig. 5.12). States with large n_r but small m (i.e. whispering gallery modes) as well as the symmetric partners of the doublets are the ones which first leave the system through the opening. This is confirmed by measurements of field intensity distributions (see Fig. 5.5) which are in good agreement with the graphical results of [79, 80] where a semicircular cavity has been attached to two leads.

6 Double-slit experiments with microwave billiards

6.1 Motivation

The diffraction of an incident plane wave by a single slit in a two dimensional region is well understood. There exist approximations to the solution of this problem depending on the ratio of the wavelength λ to the slit size d and on the distances between the source, the slit plane and the observation point. Some examples come from optics (the far field Fraunhofer approximation [96], the near field diffraction of Fresnel [97], the Fresnel-Kirchhoff diffraction integral [97] and Keller´s geometrical theory of diffraction [98]). But there is one exact method for solving this problem: the so called Fourier-Lamé method [99], which is based on the separation of the wave equation in elliptical coordinates.

For the double-slit problem, however, no exact analytical solution is known. In 1801 Thomas Young performed for the first time the double slit experiment [36]. Light beams passing through a double slit produced interference fringes. Assuming the incidence of a plane wave, the interference pattern at a distance much larger than the separation between slits is well described by the Fraunhofer diffraction integral [97].

In 2005, G. Casati and T. Prosen performed a numerical simulation of a double-slit experiment with a Gaussian wave packet which is initially confined inside a billiard domain [43]. This domain is connected to an exterior region via two small slits in a straight wall of the billiard. The resulting interferences after successive diffractions of the packet are detected in the exterior region. Both regions are two-dimensional. The evolution of the wave packet is governed by the time-dependent Schrödinger equation. Two different shapes for the boundary of the billiard domain are used. A $\pi/4$ triangular billiard represents a completely integrable system (see left picture of Fig. 6.1). If the hypotenuse of the triangle is replaced by a circular arc the system becomes fully chaotic (see right picture of Fig. 6.1).

As time proceeds the wave packet bounces at the billiard walls and once it encounters the slits, part of the wave packet exits the billiard. For the study of

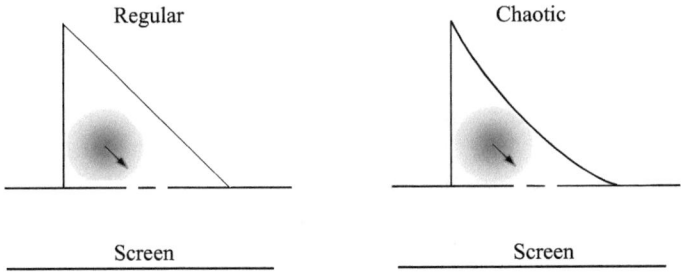

Fig. 6.1: Sketch of the open quantum billiards used in the numerical simulation of [43]. The initial state is taken as a Gaussian wave packet pointing to the slits (symbolized with a cloud and an arrow).

the interference patterns in the exterior region the probability current

$$\vec{j}(\vec{r},t) = \frac{\hbar}{2mi}(\vec{\nabla}\psi(\vec{r},t)\psi^*(\vec{r},t) - \vec{\nabla}\psi^*(\vec{r},t)\psi(\vec{r},t)) \tag{6.1}$$

is recorded along a horizontal line parallel to the straight wall of the billiard with the slits at a distance l (in the following referred to as the screen, see Fig. 6.1). It should be remarked that, in contrast to the typical double-slit experiment performed in a stationary regime, i.e. at a certain frequency, the present case deals with time-dependent interference patterns. The main result of this calculation is shown in Fig. 6.2. The time- averaged component perpendicular to the screen,

$$I(x) = \int dt\, j_y(x, y = l, t), \tag{6.2}$$

is plotted versus the distance to the middle point on the screen corresponding to the center point between the slits. In the case of the billiard with regular dynamics, the interference pattern is similar to the well known result from double-slit

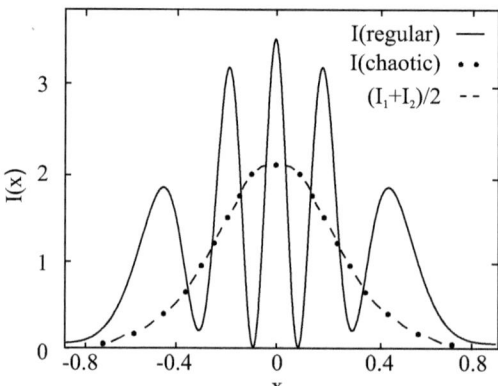

Fig. 6.2: Main result of the numerical simulation presented in [43]. The solid line shows the intensity on the screen for the double-slit experiment in the regular billiard. The dotted line is the intensity for the double-slit experiment in the chaotic billiard and the dashed line represents the sum of the intensities of the corresponding single-slit experiments. The position on the screen is given in units of the billiard length.

experiments with plane waves in the Fraunhofer regime (far field region), i.e. the encountered intensity equals the sum of the intensities from the two corresponding single-slit experiments plus an interference term. In the case of the billiard with chaotic dynamics, the intensity on the screen turns out to equal the superposition of the two diffraction maxima radiated from each of the single slits. As a consequence the fringes disappear and the intensity becomes unimodal. Note that the visibility of the fringes depends on the initial parameters of the wave packet and the clear distinction between the regular and chaotic dynamics observed in Fig. 6.2 was only obtained for certain initial conditions [100].

The experiments described in the next sections were initially motivated by these results. However, in experiments with microwave billiards the time evolution of an initial state is described by the electromagnetic wave equation instead of the time-dependent Schrödinger equation. Patterns obtained from the emission from single and multiple antennas are numerically and experimentally studied. The emission from a single antenna provides an omnidirectional wave pulse. We, moreover, developed a new method, where the microwave power was excited by arrays of multiple antennas in order to obtain a directed initial state as applied in the numerical simulation of [43]. This represents a novel method in the investigation with microwave billiards [101].

6.2 Experiment

In the experiments two microwave billiards with different shapes have been used, a rectangular one for studying interference patterns from a system with regular dynamics and a desymmetrized tilted stadium for a system with chaotic dynamics. The upper panel of Fig. 6.3 shows a photography of the top view of the rectangular cavity with dimensions 768 × 475 mm. The ratio of the side lengths is taken as the golden ratio $(\sqrt{5}+1)/2$ in order to avoid degeneracies of the eigenmodes. The lower panel of Fig. 6.3 shows a photography of the cavity with the shape of a desymmetrized tilted stadium. The radius of the quarter circle is 415 mm and equals the length of the bottom side of the attached trapezoid. The length of the left side is 370 mm. Both billiards enclose the same area and thus provide

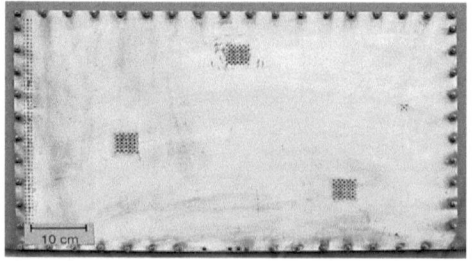
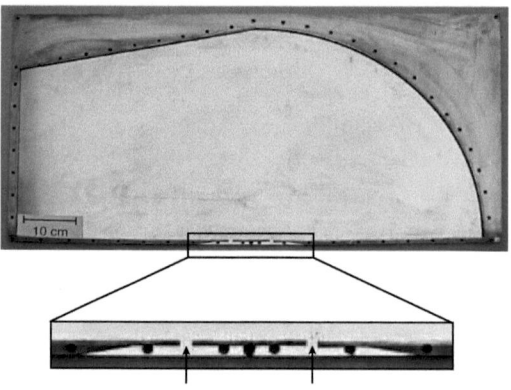

Fig. 6.3: Photographs of the microwave cavities used in the double-slit experiments. The top view of the rectangular billiard is shown in the upper panel. The shape is defined by the screws which squeeze together the bottom and top plate and the frame in between. In the middle panel the billiard in form of a tilted stadium is shown. The slits are indicated by arrows in the zoom of the slit region.

similar numbers of states according to Weyl´s formula [93]. The construction of the microwave cavities is modular as described in Sec. 4.1. The bottom side of the billiards shown in Fig. 6.3 is composed of three modular bars of copper which enable the variation of the slit size and distance between slits (d and s in Fig. 6.4, respectively). The slits are magnified in the lower photography of Fig. 6.3 (the circular holes visible in the zoom are used to screw the copper bars to the top and bottom plates).

Sets of small holes (2 mm diameter) are drilled in the top plates of the billiards for the microwave antennas (see upper photography of Fig. 6.3). Their purpose will be explained in Secs. 6.5 and 6.6.

Figure 6.4 illustrates schematically the experimental setup. The RF power, generated by the VNA, is coupled into the billiard through a fixed antenna a placed at a fixed position inside the billiard (see Sec. 4.1). The wire of the antenna a penetrates 2.5 mm, i.e. half the height of the cavity, into the billiard. In order to measure the electrical field due to the leakage of microwave power through the slits, a receiving antenna b of 15 mm length is moved in steps of 5 mm (one third of the minimal wavelength) over a grid of points closely lying to the billiard edge. For this antenna a flexible coaxial cable is used. The distance l between the edge of the billiard and the closest measuring point is set to 20 mm. The middle point

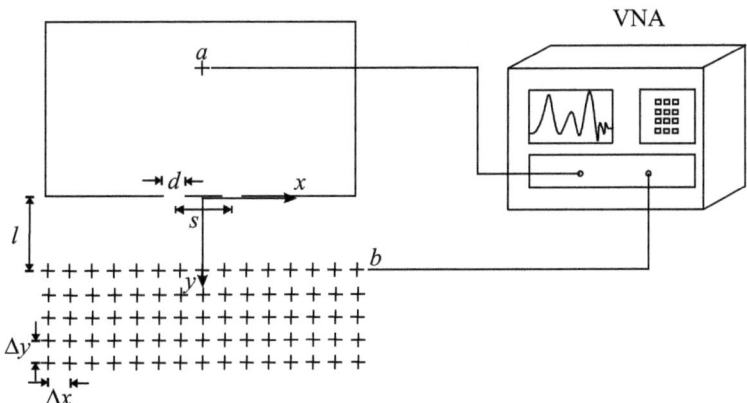

Fig. 6.4: Schematic view of the experimental setup. The antenna a is fixed inside the billiard and the antenna b is moved in small steps ($\Delta x = \Delta y = 5$ mm) in the vicinity of the slits to measure the electrical field. The distance between the slits and the slit size are denoted by s and d, respectively. The distance to the measuring points is denoted by l. The coordinate system (x, y) used in the following is depicted. The microwave radiation is generated by the VNA.

of the metallic pin of the antenna b is carefully aligned with respect to the plane defined by the billiard slits.

The antenna b is guided by a carriage attached to a positioning unit (see upper photography of Fig. 6.5). This unit is driven by bipolar step motors and connected to a computerized numerical control (CNC) module. A PC programm communicates via serial interface with the VNA and the CNC. The VNA is calibrated in order to remove the effects of the cables and the connectors on the measured spectra. These effects include reflections at connectors, attenuation in the cables and the time delay of the signal corresponding to the length of the cables (see Sec. 4.1). The transmission element S_{ba} of the scattering matrix is measured between the antenna a and b. The transmission amplitude can be related to the product $E_z(r_a)E_z(r_b)$ of the electric fields enabling the access to the electrical field $E_z(r_b)$ outside the cavity (see Sec. 4.1). For each position of the antenna b, a frequency spectrum from 0.5 to 20 GHz with a step of 5 or 10 MHz is taken. Such a large frequency step (cf. with measurements presented in Sec. 5.2.3 where the step was 15 kHz) is used since the measurements with a movable receiving antenna are time-consuming.

The measurements were carried out inside a so called anechoic chamber (see the lower photography of Fig. 6.5). Such a room is used by microwave engineers for far field RF measurements. Pyramidal polyurethane foam structures VHP-12 NRL (from the company Emerson & Cuming) cover the whole surface of the chamber and ensure microwave attenuation of the reflected power down to -50 dB over the frequency range, in which the measurements are performed [102].

All metallic parts of the measurement are covered with microwave absorption material. The absorption material consists of urethan foam sheets impregnated with carbon (EPP-51 material from the company ARC Technologies). Moreover, the cable connected to the receiving antenna is coated by an absorptive double layer (urethan foam EPP-51 and urethan flat plate EPF-11) in order to suppress reflections between the RF-cable coat of copper and the billiard edge.

Figure 6.6 shows a frequency spectrum for the rectangular billiard taken with the emitting antenna a placed in the resonator and the receiving antenna outside at the point $x=0$, $y = 320$ mm (see coordinate system introduced in Fig. 6.4). The antenna a is put symmetrically with respect to the slits at a distance 400 mm from

Fig. 6.5: Experimental setup for the measurement of interference patterns (upper photography). The microwave billiard is placed on a table and carefully aligned with the moving antenna, which is coated with an absorber foam and guided by a positioning unit. The microwave power is coupled from the vectorial network analyzer through a coaxial cable into the cavity. The lower photography shows a full view of the anechoic chamber.

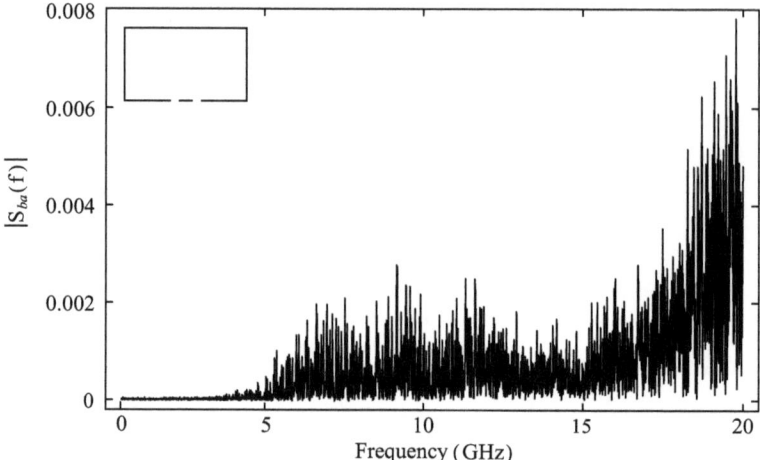

Fig. 6.6: Frequency spectrum of the open two-slit rectangular billiard measured with the emitting antenna a inside the resonator and the receiving antenna b placed outside.

the lower billiard edge. The slit size d is set to 9.5 mm and the distance s between slits is 78 mm. The spectrum shows many sharp resonances corresponding to quasi-bound modes. Below the cut-off frequency $c/2d = 15$ GHz defined by the slit size (see Sec. 5.2.2), the transmission amplitude is notably suppresed[5]. Below 5 GHz the signal is completely attenuated. The narrow transmission amplitude can be realized in the small values on the vertical axis.

6.3 Stationary wave patterns

The squared modulus of the measured transmission scattering matrix element $|S_{ba}|^2$ as function of the position x of the receiving antenna b is presented in

---

[5]In contrast to the results of Fig. 5.6 described in Sec. 5.2.2 propagation of modes for frequencies smaller than the cutoff frequency of the slits is observed. Due to their small width the slits cannot be treated as a lead with parallel walls.

Fig. 6.7 (solid curves) for the rectangular (left panel) and the stadium billiard (right panel). It is measured at a distance $l=155$ mm from the bottom edge of the billiard (for definition of l see Fig. 6.4). The frequencies are chosen as resonance frequencies of the corresponding billiards. The slit size and the distance between slits are $d=8$ mm and $s=76$ mm, respectively. The solid curves in Fig. 6.7 correspond to a symmetrical position of the emitting antenna with respect to the slits, i.e. $x=0$. For the rectangular billiard, we observe an interference pattern with a central maximum located at $x=0$ and two symmetrical lateral maxima. If the emitting antenna is placed away from the line $x=0$ through the middle point between the slits, we obtain a very similar interference pattern i.e. the measured patterns are nearly independent of the position of the emitting antenna as expected from a stationary field distribution inside the resonator. Similar results from double-slit experiments with water surface waves from billiards with regular dynamics have been published in [103]. It should be noted that the motion of water waves also obeys the two dimensional Helmholtz equation and their motion in billiards with hard walls is equivalent to a quantum mechanical problem of a particle moving inside such a billiard.

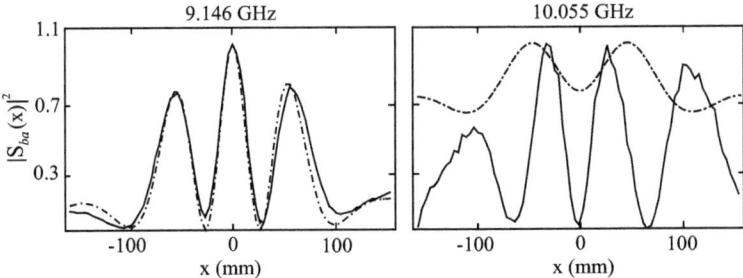

Fig. 6.7: Stationary intensity patterns for the rectangular billiard (left panel) and the stadium billiard (right panel). The solid lines correspond to a position of the emitting antenna on the line $x=0$. The theoretical patterns corresponding to Eqs. (6.3) and (6.6) are plotted as dot-dashed lines in the left and right panel, respectively.

The interference patterns for the rectangular billiard are well described by the Fraunhofer formula,

$$I(x) = I_1(x) + I_2(x) + 2\sqrt{I_1 I_2} \cos(\frac{ksx}{l}). \tag{6.3}$$

which is valid in the far field region with $d^2/(l\lambda) < 1$ [97]. For the parameters of the setup presented in Fig. 6.7, $d^2/(l\lambda) \approx 0.01$. The formula consists of the sum of the intensities $I_1(x)$ and $I_2(x)$ radiated by each individual slit plus an oscillatory term which describes the interferences between the emitted waves by the slits [97]. The intensities I_1 and I_2 are obtained from the corresponding single slit experiments and are well described by $I(x) = \text{sinc}(\frac{kdx}{l})^2$. The curve corresponding to Eq. (6.3) is plotted as dot-dashed line in the left panel of Fig. 6.7.

According to the so called random plane wave model [5], the wave function in a chaotic billiard can be described as a superposition of plane waves with the same wave number $|\vec{k}_n| = k$ and random amplitudes and directions, i.e.

$$\psi_k(\vec{r}) = \sum_n a_n e^{i\vec{k}_n \vec{r}}. \tag{6.4}$$

This ansatz yields a Bessel function of the first kind of order 0 for the spatial correlator [5]

$$\langle \psi_k(\vec{r}) \psi_k(\vec{r'}) \rangle = J_0(k|\vec{r} - \vec{r'}|). \tag{6.5}$$

This implies that the fields at the position of the slits are correlated as $J_0(ks)$ and the interference term of Eq. (6.3) is modified [103]. Accordingly

$$I = I_1 + I_2 + 2\sqrt{I_1 I_2} J_0(ks) \cos(\frac{ksx}{l}). \tag{6.6}$$

For wave numbers k for which $J_0(ks) = 0$ the interference term vanishes and the intensity should equal the sum of intensities of single-slit experiments. A measured interference pattern for the tilted stadium is shown as solid line in the right panel of Fig. 6.7. The intensity corresponding to Eq. (6.6) is plotted as dot-dashed line. No agreement between Eq. (6.6) and the experimental results is found. In fact, the visibility $(I_{\max} - I_{\min})/(I_{\max} + I_{\min})$ of the measured pattern is comparable with the one of patterns of the rectangular billiard. However, in general the patterns do not exhibit spatial symmetry as in the rectangular billiard. Indeed, for all stationary patterns that were measured the interferences never disappear, neither for wave numbers canceling the interference term, i.e. $J_0(ks) = 0$, nor for other k. It should be noted that also the results for chaotic billiards with water surface waves [103] do not provide convincing evidence that the total intensity equals the sum of the intensities for the single-slit experiments.

Examples of two dimensional intensity patterns for the rectangular and stadium billiard are shown in Fig. 6.8. The slit size d is 20 mm and the distance between slits s is set to 240 mm. The grid of measured points covers an area of 0.5 m^2 (see Fig. 6.8). In the two upper left panels the intensity patterns for two isolated resonances of the rectangular billiard with frequencies f=5.758 GHz and 7.731 GHz are shown. The emission patterns are symmetric with respect to the line x=0. This is due to the symmetrical distributions of intensities inside the resonator. In the regime of overlapping resonances the transmission spectrum shows an oscillating structure. The frequencies $f = 10.910$ GHz and 17.025 GHz correspond to two maxima of these oscillations. In this case the interference patterns are slightly asymmetric. The asymmetry is related to the overlapping of the resonances. Consequently, several modes are excited simultaneously with different strength such that the resulting field distribution has no spatial symmetry. For the stadium billiard non-symmetric interference patterns are observed reflecting the random field distribution of the corresponding modes. There are even cases where the emission mainly takes place from just one slit, for instance for $f = 4.557$ GHz and for $f = 10.124$ GHz in Fig. 6.8.

These results provide the first insight into the effect of the shape of the billiards on the interference patterns. For the regular case patterns similar to those obtained in the original double-slit experiments are obtained. It should be remarked that such interference patterns are only observed for sharp resonances. For the stadium billiard, interferences are also observed in the stationary intensity patterns. Their strong asymmetry with respect to the line x=0 is due to the spatial distributions of the chaotic eigenstates. In the next sections the time evolution of interference patterns is investigated.

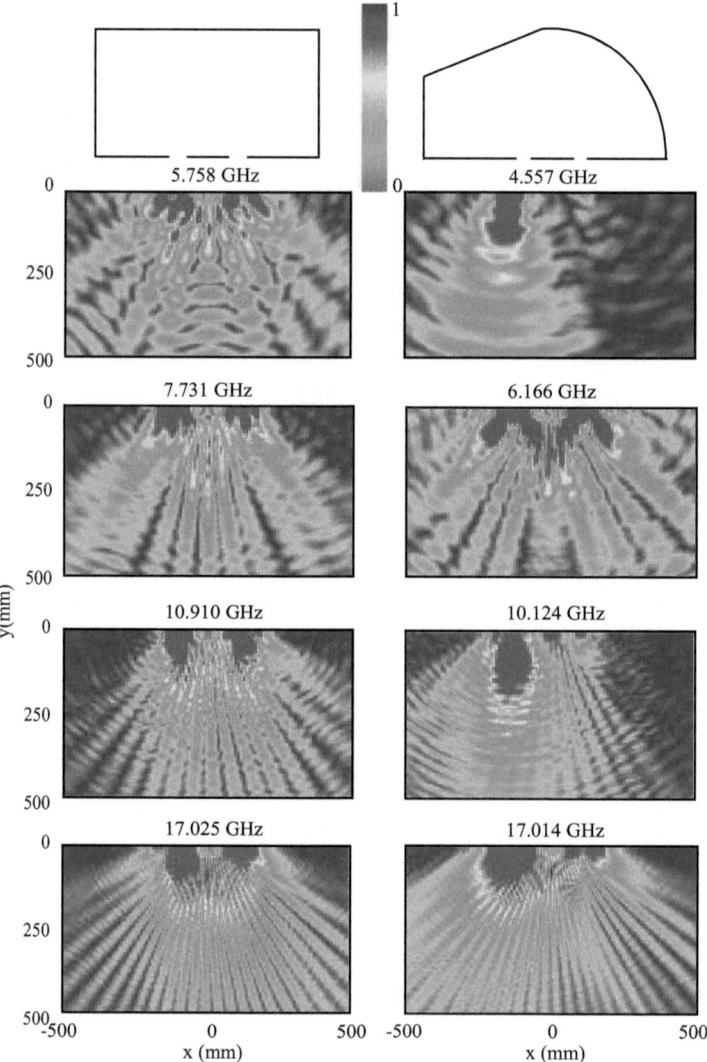

Fig. 6.8: Two dimensional intensity patterns over a grid of measuring points for the rectangular (left panels) and tilted stadium billiard (right panels).

6.4 Interference patterns in the time domain

6.4.1 Time-dependent Schrödinger equation versus Helmholtz equation

In quantum mechanics the time evolution is governed by the time-dependent Schrödinger equation

$$i\hbar \partial_t \psi(\vec{r},t) = \hat{H} \psi(\vec{r},t) \tag{6.7}$$

where $\hat{H}$ is the Hamilton operator. In the simulation of [43] described in Sec. 6.1, the initial state is taken as a Gaussian wave packet

$$\psi(\vec{r},0) = \exp\left(-\frac{(\vec{r}-\vec{r}_0)^2}{2\sigma^2}\right) \exp\left(i\vec{k}(\vec{r}-\vec{r}_0)\right) \tag{6.8}$$

propagating in the direction of the wave vector $\vec{k}$. By means of the quantum mechanical propagator $K(\vec{r},\vec{r}_0,t,t_0)$, the time evolution of an initial state $\psi(\vec{r}_0,t_0)$ can be expressed as [104]

$$\psi(\vec{r},t) = \int d\vec{r}_0 K(\vec{r},\vec{r}_0,t,t_0) \psi(\vec{r}_0,t_0). \tag{6.9}$$

The evolution of electromagnetic waves, which is relevant for the experiments presented here, is described by the differential equation

$$\frac{1}{c^2} \partial_t^2 E_z(\vec{r},t) = \Delta E_z(\vec{r},t). \tag{6.10}$$

The time evolution of the initial state in the electromagnetic case can also be expressed with the propagator formalism

$$E_z(\vec{r},t) = \frac{1}{4\pi c^2}\left[\int_{\mathcal{G}} d^2\vec{r}_0 K(\vec{r},\vec{r}_0,t,t_0) \frac{\partial E_z(\vec{r}_0,t_0)}{\partial t} - \int_{\mathcal{G}} d^2\vec{r}_0 \frac{\partial K(\vec{r},\vec{r}_0,t,t_0)}{\partial t} E_z(\vec{r}_0,t_0)\right], \tag{6.11}$$

where $K(\vec{r},\vec{r}_0,t,t_0)$ denotes the electromagnetic propagator [105] and $\mathcal{G}$ is the bounded domain from which the initial state $E_z(\vec{r}_0,t_0)$ is evolved (in the present case the billiard surface). As can be seen from the above considerations, the wave equations describing the time evolution of wave packets in microwave and quantum billiards are not equivalent. The situation differs clearly from that of the stationary case where one can exploit the analogy of the Schrödinger equation and the scalar Helmholtz equation (see Sec. 2).

6.4.2 Emission from a single antenna in free space

In Sec. (4.1) a relation of the measured transmission amplitude S_{ba} between antennas at positions r_b and r_a and the cavity Green function $\hat{G}(\vec{r}_b, \vec{r}_a, f)$ was shortly outlined. In analogy the propagator is related to the Fourier transform of the experimentally obtained scattering matrix transmission element, i.e.

$$\tilde{S}_{ba}(t) = \int_{-\infty}^{\infty} S_{ba}(f) e^{-2\pi i t f} df \sim K(\vec{r}_a, \vec{r}_b, t). \tag{6.12}$$

The experimental data consist of a set of N data points from $f_{\min}$ to $f_{\max}$ at discrete frequencies and correspondingly a discrete Fourier transform has to be applied to the measured data,

$$\tilde{S}_{ba}(t_j) = \Delta f \sum_{n=0}^{N-1} S_{ba}(f_n) e^{-i 2\pi t_j f_n} \tag{6.13}$$

where $t_j = (j-1)/(f_{\max} - f_{\min})$ and Δf is the experimental frequency step (5 MHz).

To illustrate the relation between $\tilde{S}_{ba}(t)$ and $K(\vec{r}_a, \vec{r}_b, t)$ we performed a measurement of the transmission between two antennas in free space. A fixed emitting antenna a is hanging from the ceiling of an empty room and the receiving antenna b is moved around in a plane perpendicular to the antenna wire a in steps of 5 mm. The minimal and maximal frequencies are set to 0.5 and 20 GHz, respectively. The grid of measured points, i.e. positions of the antenna b covers a surface of 1000×600 mm. Thus the S_{ba} elements are measured as function of the spatial coordinates and frequency. The propagation of the wave pulse emitted from the fixed antenna in all directions is clearly observed in the three snapshots for $t=0.2$, 0.4 and 1 ns (top to bottom) shown in Fig. 6.9.

In free space the theoretical expression for the propagator introduced in Eq. (6.11) is obtained as [105]

$$K(\vec{r}, \vec{r}_0, t, t_0) = \frac{\Theta(t - t_0 - |\vec{r} - \vec{r}_0|/c)}{\sqrt{(t - t_0)^2 - |\vec{r} - \vec{r}_0|^2 / c^2}} \tag{6.14}$$

where $\Theta(x - x_0)$ denotes the step function, i.e. $\Theta(x - x_0) = \begin{cases} 1 & \text{if } x \geq x_0 \\ 0 & \text{if } x < x_0 \end{cases}$.

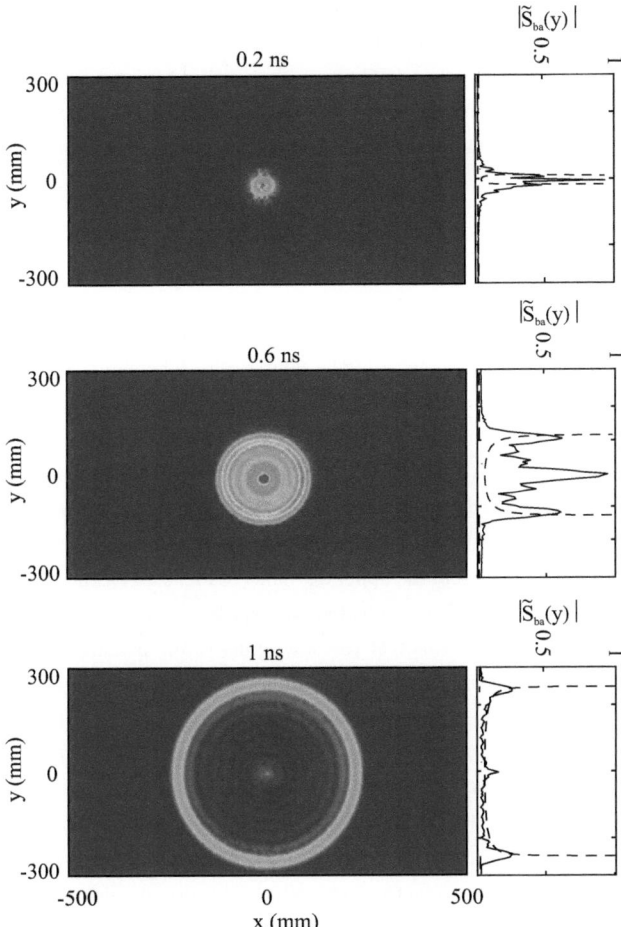

Fig. 6.9: Three snapshots of the propagation of the wave pulse emitted by a dipolar antenna. For the color scale see Fig. 6.8. The measurement was performed in free space. The right panels show the propagation projected along the line $x=0$ as solid line. The propagator corresponding to Eq. (6.14) is plotted as dashed line.

The solid lines in the plots shown in the right panels of Fig. 6.9 result from the projection of $|\tilde{S}_{ba}(\vec{r}_b = (0, y), t)|$. The propagator corresponding to Eq. (6.14) is plotted as dashed line. The behaviour of the measured wave pulse is well described by the theoretical curve. The central peak and the inner ring structure observed in the measurements is not well described by the theoretical propagator. They correspond to a quasi-bound mode trapped between the antennas.

6.4.3 Emission from a single antenna in microwave billiards

In this section we analyze the spatial structure and the time evolution of the field outside the billiards in the vicinity of the double slit. The emitting antenna a is placed inside the billiard symmetrically with respect to the slits at a distance of 400 mm to the slits whereas the distance l of the receiving anntena b to the edge of the billiard equals 320 mm (see Fig. 6.4). The antenna b is also located symmetrically with respect to the slits. The slit size and the distance between the slits are 8 mm and 78 mm, respectively. The squared modulus of the Fourier transform $|\tilde{S}_{ba}|^2$ is presented in the upper left panel of Fig. 6.10 for the rectangular and in the upper right panel for the tilted stadium billiard. The position of the peaks correspond to travel times of the wave pulse after passing through the slits, i.e. classical paths from antenna a to antenna b. Some examples of such classical paths are shown in the upper pictures. The positions of the peaks are called escape times in the following. The first peak in the left panel of Fig. 6.10 (rectangular billiard) has a double structure: The prominent peak (the first escape time) corresponds to the time of flight for a direct path between antennas a and b through the slits (path 1 in the upper left picture) whereas the smaller peak (the second escape time) corresponds to a path which includes an additional bounce of the wave pulse off the upper boundary (path 2 in the upper middle picture). The third escape time corresponds to the reflection of the wave pulse in the left and right boundary before reaching the slits (path 3 in the upper right picture). Following this procedure we can assign a classical path between a and b to all peaks in the upper panels of Fig. 6.10.

In addition to the experimental results plotted in the upper panels of Fig. 6.10, results from a ray tracing simulation of the classical billiards are presented in

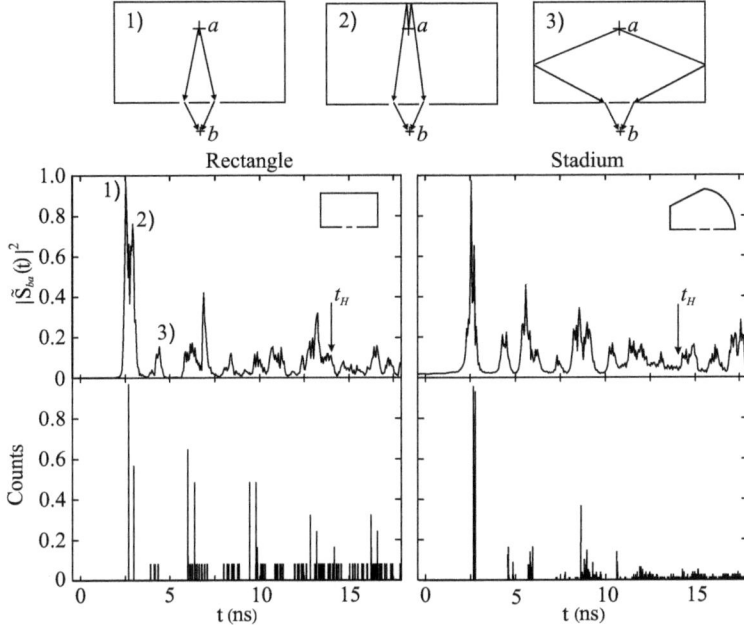

Fig. 6.10: Time spectra for the rectangular and the stadium billiard (upper panels). The antenna b is sited behind the double-slit at a distance of 320 mm. Results from a ray tracing simulation are shown in the lower panels. The Heisenberg time t_H is marked with an arrow. The upper pictures illustrate the shortest classical paths from antenna a to antenna b. For details, see the main text.

the lower panels. The simulation is done as follows: A point-like particle is injected from the position of antenna a. In order to simulate the omnidirectional character of an initial wave pulse, the initial direction of the particle sweeps $5 \cdot 10^6$ equidistant angles from 0 to 2π. The particles leave the classical billiard via the slits. Once a particle escapes, its time of flight from the slit to the position of antenna b is added to its time of flight inside the billiard and the total time of flight is recorded. The particle number corresponding to a given time of flight is counted and shown in the lower panels of Fig. 6.10. Below the Heisenberg time

t_H (equal to 13.8 ns for both systems, see Sec. 5.3.2.2), the classical times of flight agree well with the escape times of the experimental wave pulse. The quantum dynamics is expected to follow the classical behaviour for times $t < t_H$ as clearly shown in [34]. Then the good correspondence between the quantum-mechanical and classical results allows us to ascribe each peak in the quantum-mechanical transmission amplitudes to a specific classical trajectory connecting the antennas a and b.

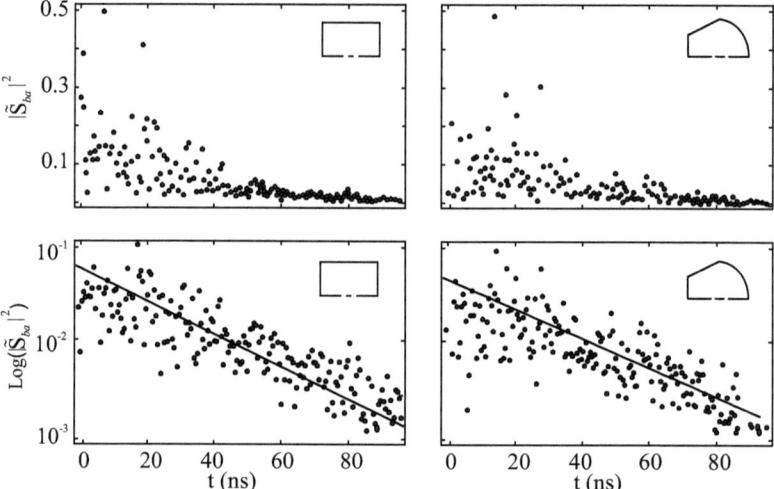

Fig. 6.11: Decay probability for the rectangular billiard (left panels) and for the stadium billiard (right panels). The probability decreases exponentially in the time window accessible in the experiments (see logarithmic plots in the lower panels).

Figure 6.11 shows the decay probability of the two billiards in linear scale (upper panels) and logharitmic scale (lower panels)[6]. The parameters d, s, l and the position of the emitting antenna a are the same as for the results

---

[6]In contrast to the measurements presented in Sec. 5.2.3, the decay probability through the slits is here mainly governed by absorption in the billiard walls (in analogy to the decays shown in the left panel of Fig. 5.7).

presented in Fig. 6.10. The least square fit of an exponential function $e^{-\beta t}$ yield decay constants $\beta = 0.043 \pm 0.003$ ns^{-1} for the rectangular billiard and $\beta = 0.036 \pm 0.005$ ns^{-1} for the stadium. Correspondingly, after approximately $t = 50$ ns the squared transmission amplitude has decayed by a factor of e^{-2}.

Figure 6.12 shows an intensity plot for the rectangular and the stadium billiard, i.e. the squared modulus Fourier transform of the scattering matrix element $|\tilde{S}_{ba}|^2$ versus the coordinate x of the antenna b along the horizontal axis and the time t along the vertical axis. The parameters d, s, l and the position of the emitting antenna a are the same as for the results presented in Fig. 6.10.

For every escape time, i.e. every time the diffracted wave pulse arrives at the positions of the antenna b, an interference pattern is observed as a color variation along the x axis. This pattern is due to interferences between the waves emitted from the two slits when the wave pulse hits them. As a result of the spatial symmetry of the rectangular billiard, every element of the cylindrical wave pulse emitted by the antenna with initial momentum direction $\vec{k} = (k_x, k_y)$ travels the

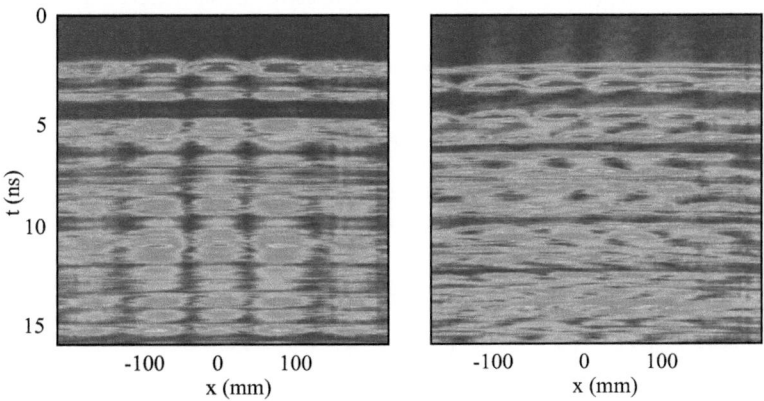

Fig. 6.12: Time evolution of the intensity patterns of the rectangular (left panel) and stadium billiard (right panel). The squared modulus of the Fourier transform $|\tilde{S}_{ba}|^2$ is represented versus the coordinate x along the horizontal axis and the time t along the vertical axis. For the color scale see Fig. 6.8.

same path as its counterpart with $\vec{k} = (-k_x, k_y)$. Accordingly, when they hit the slits there is no phase difference between them and the maxima and minima of their interferences remain at the same positions independent of the escape times. For the stadium billiard, however, there is no spatial symmetry and consequently the positions of the maxima and minima move to the right and to the left.

In order to quantify how the interference patterns are affected by the time evolution of the phase difference between the waves emitted from the two slits, the measured intensities $|\tilde{S}_{ba}(t)|^2$ are averaged over time. It should be noted that this is equivalent to evaluating Eq. (6.2). The maximal time experimentally accessible corresponds to $\frac{1}{\Delta f}$, where Δf is the frequency step size used in the experiment. In the present measurements $\Delta f = 10$ MHz which corresponds to a maximal time of 100 ns. The time average is performed up to 75 ns (which corresponds to about five times the Heisenberg time) since the contributions of $|\tilde{S}_{ba}(t)|^2$ at longer times are negligible small (see Fig. 6.11).

The resulting time averaged intensity patterns (see Fig. 6.12) are shown in Fig. 6.13 as solid lines. For the rectangular billiard, the interferences survive the time averaging. For the stadium billiard, however, the interference fringes disappear as a result of the shiftings of the maxima and minima as function of time. Small remnants of an interference pattern are due to dominant peaks coming from the first two escape times (see Fig. 6.12).

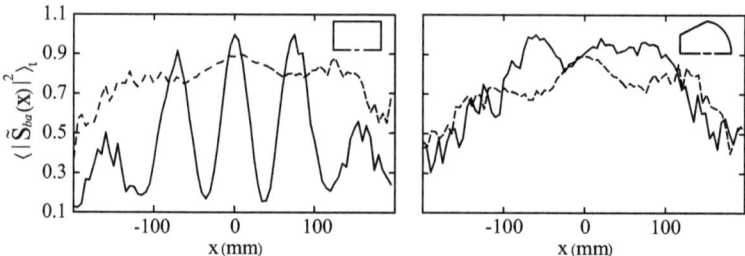

Fig. 6.13: Comparison of the time-averaged intensity from double-slit experiments (solid lines) with the sum of the intensities from the corresponding single-slit experiments (dashed lines). The left panel corresponds to the rectangular billiard and the right panel to the stadium billiard.

Beside double-slit experiments, single-slit ones are also performed. In Fig. 6.13 the sum of the squared modulus $|\tilde{S}_{ba}|^2$ measured in two single-slit experiments (dashed lines) is compared with the result of the double-slit experiment. For the rectangular billiard the interference fringes are only present in the case of the double-slit. For the stadium billiard, however, the shapes of both curves overall agree reproducing the expected diffraction maximum.

As explained above, in the case that the emitting antenna a is placed inside the rectangular billiard on the symmetry line between the slits $x = 0$, the phase difference of the waves emitted by the two slits does not change with time and thus we observe interferences. However, if the emitting antenna is displaced from the symmetry line between slits, the interference fringes of the time-averaged intensity distribution disappear (see Fig. 6.14). This can be understood since the waves exiting the two slits do not travel the same path length and thus have a varying phase relation. Thus the result is more similar to that from the stadium cavity as can be seen in the right panel of Fig. 6.13. This finding confirms that the mechanism leading to the emergence of the interference fringes observed in the left panel of Fig. 6.13 is linked to the position of the emitting antenna with respect to the slits, i.e. they show up as a consequence of the spatial symmetry

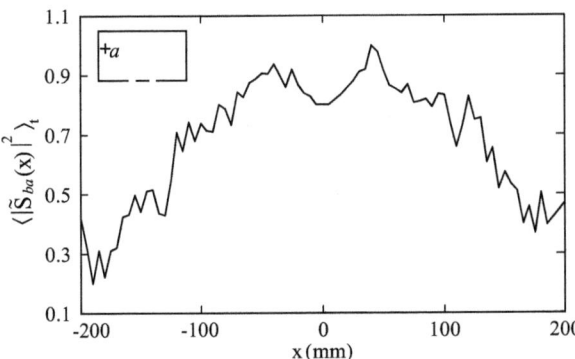

Fig. 6.14: Intensity measured outside the rectangular billiard with two slits. The emitting antenna is positioned non symmetrical with respect to the slits and located at a distance of 237.5 mm from the lower edge of the billiard and 10 mm from the left one as depicted in the inset.

and not of the dynamics. The results of the experiments with a single emitting antenna presented in this section lead to the following conclusions:

- Clear interference fringes are only observed for the rectangular billiard with the emitting antenna placed on the symmetry line $x=0$ defined by the slit positions.

- Interference fringes are not observed for the stadium billiard and for the rectangular billiard with emitting antenna placed asymmetrically with respect to the slits.

- For the stadium billiard the sum of the intensities of the corresponding single-slit experiments is roughly equal to the intensity of the double-slit experiment.

As explained in Sec. 6.1, the initial state of the numerical simulation of [43] corresponds to a localized wave packet with a well defined direction in clear contrast to the propagation shown in Fig. 6.9. Thus the experimental results presented in this section are not directly comparable to the simulation. To further explore the differences in the intensity patterns from the billiards with regular and chaotic dynamics, the next sections deal with the preparation of initial states with a well defined direction.

6.5 Excitation of the billiards with an elongated wave packet

6.5.1 Preparation of an elongated wave packet

As explained in Sec. 6.4.3 the experiments with a single emitting antenna do not have the same initial state as in the numerical simulation of Casati and Prosen where a wave packet with a given propagation direction was considered [43]. One method how to create an initial state with a preferred direction comes from the field of radar engineering. Let us consider a linear array of antennas. If all antennas emit in phase the circular wave fronts evolving from every single antenna will be superimposed to form two elongated wave packets similar to plane waves propagating perpendicular to the array line, one moving forward and another one backward. Moreover, if consecutive antennas emit with a certain constant phase difference, the resulting elongated wave packet will be then inclined with respect to the normal to the array line. This is illustrated in the left panel of

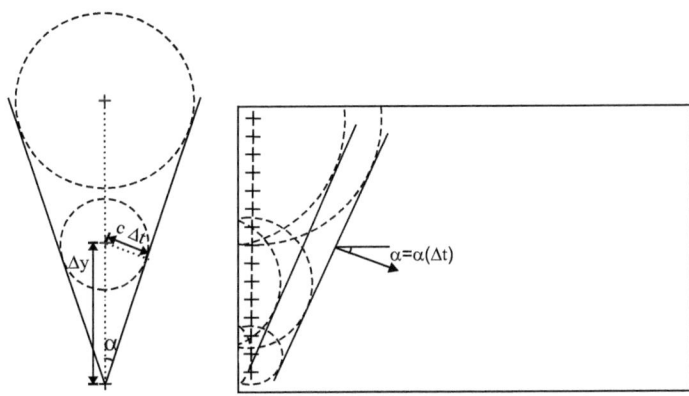

Fig. 6.15: Schematic picture demonstrating the procedure for the excitation of an elongated wave packet using an array of equidistant antennas positioned close to the boundary.

Fig. 6.15, where the wave fronts emitted by each antenna are depicted as circles. The signals of neighbouring antennas at a distance Δy are emitted with a time delay Δt. Accordingly the initial angle α of the elongated wave packet is

$$\alpha = \arcsin \frac{c\Delta t}{\Delta y}. \tag{6.15}$$

As can be seen from Eq. (6.15), the proposed method offers the advantage of varying the initial angle α of the elongated wave packet by using different time delays Δt. The resulting elongated wave fronts from the linear array of antennas are depicted as solid lines in Fig. 6.15. The backward front is reflected at the left boundary into the same direction as the forward front. It is important to place the antennas close to the boundary so that the forward and backward wave fronts are as close to each other as possible. On the other hand close to the boundary the field intensity decreases due to the Dirichlet boundary condition (2.2). It was checked numerically that the most effective excitation of the elongated wave packet is obtained for an array of antennas 10 mm apart from the left boundary.

For the measurements a vertical array of holes for the emitting antennas has been drilled close to the left boundary of the billiards into the top plate of the microwave billiards (see right panel of Fig. 6.15). In the experiments a frequency spectrum is measured for each emitting antenna of the array separately and subsequently Fourier transformed as explained in Sec. 6.4.2. Because of the linearity of Eq. (6.10), we can superpose the Fourier spectra of every antenna. Thus, the electric field of the resulting elongated wave packet can be written as

$$E_z(\vec{r},t) \propto \sum_i \left(\int_{\mathcal{G}} d^2\vec{r}_0 K(\vec{r}_b, \vec{r}_{0i}, t, t_0+i\Delta t) \frac{\partial E_z(\vec{r}_{0i}, t_0)}{\partial t} \right) \propto \sum_i \tilde{S}_{ba}(\vec{r}_b, \vec{r}_{0i}, t_0+i\Delta t), \quad (6.16)$$

where $K(\vec{r}_b, \vec{r}_{0i}, t, t_0 + i\Delta t)$ stands for the electromagnetic propagator introduced in Sec. 6.4.1 and r_{0i} for the position of antenna i in the vertical array. The second term in Eq. (6.11) has been set to zero by proper choice of initial conditions, i.e. $E_z(\vec{r}_0, t_0) = 0$. So, we add the time spectra $\tilde{S}_{ba}$ for the different exciting antenas in the array, and each time spectrum is shifted by an amount Δt.

We first study numerically the propagation of the elongated wave packet inside the closed rectangular cavity. The propagator of Eq. (6.11) is expanded in terms of the set of eigenfunctions of the rectangular billiard as

$$K(\vec{r}, \vec{r}_0, t, t_0) \propto \sum_{n=1}^{N} \frac{\psi_n^*(\vec{r})\psi_n(\vec{r}_0)}{\omega_n} \sin(\omega_n(t - t_0)), \quad (6.17)$$

where ω_n and $\psi_n(\vec{r})$ are the resonance angular frequency and the eigenfunction of the nth state of the rectangular billiard, respectively. The number of eigenstates N used in the simulation is 4900. Three snapshots corresponding to the results of the numerical calculation are shown in Fig. 6.16. The simulation shows well directed propagation until 30 ns which is about half of the maximal time considered in the experiments (see Sec. 6.4.3). Since in the experiments the slit dimensions are small compared to the size of the billiards we may conclude that the wave

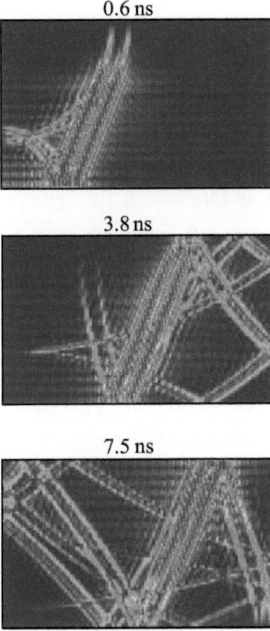

Fig. 6.16: Three snapshots corresponding to the results of the numerical simulation of the propagation of an elongated wave packet inside the rectangular billiard. An array of 59 antennas close to the left boundary has been used. For the color scale see Fig. 6.8.

packet is barely distorted due to the openings. Thus the numerical results can approximate the experimental situation.

6.5.2 Experiments with an elongated wave packet

According to the results of the numerical simulation an array of 59 antennas placed 10 mm apart from the left boundary in the rectangular billiard yields a well directed elongated wave packet. Thus we used this configuration in the experiments. The distance between neighbouring antennas Δy is 7.5 mm. The time resolution is 0.05 ns thus yielding a minimal angle α of about 66° (see

Eq. (6.15)). However, the time resolution can be increased by interpolating the discrete experimental data points in the time spectra enabling the generation of angles from 0° to 90°. The elongated wave packet is sent towards the slits and the initial angles are chosen such that the wave packet populates a certain family of periodic orbits of the rectangular billiard. The intensity of the waves that escape through the slits as function of the time of flight (in length units) is plotted in the left panels of Fig. 6.17 for angles $\alpha = 35°$ (upper panel) and $\alpha = 54°$ (lower panel) which correspond to the family of periodic orbits depicted in the insets. For $\alpha = 35°$ we observe a clear periodic structure of the escape times which we interpret as follows: The elongated wave packet is emitted from the array of antennas. At the time it hits the double slit, a part of it escapes and travels to the antenna b outside the billiard. This corresponds to the first peak in the time spectrum. The reflected part of the wave packet follows the classical

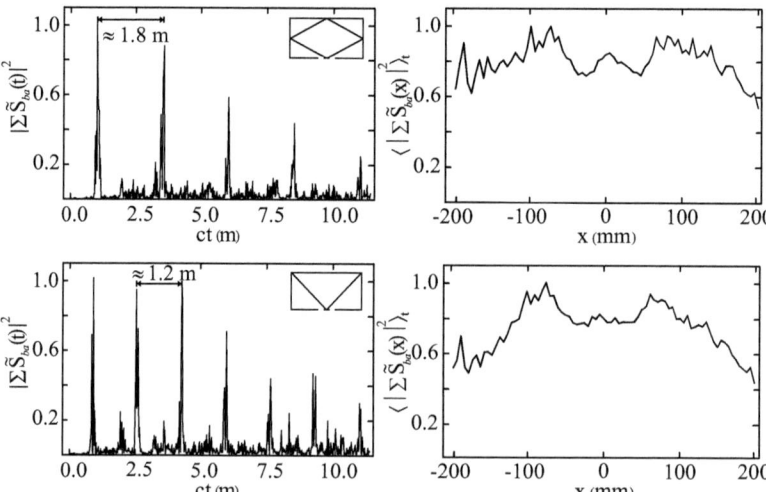

Fig. 6.17: Time spectra (left panels) and time-averaged intensity patterns (right panels) for the elongated wave packet prepared in the rectangular billiard with an array of 59 antennas. In the upper panels the initial angle is set to 35° (diamond orbit) and in the lower ones to 54° (triangular orbit).

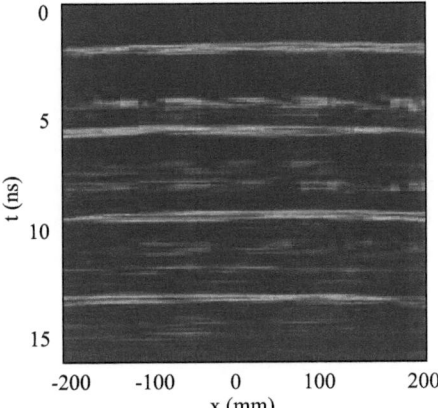

Fig. 6.18: Time evolution of the intensity pattern outside the rectangular billiard. An elongated wave packet is prepared with an angle of 54 ° (triangular orbit of Fig. 6.17). The squared modulus Fourier transform $\tilde{S}_{ba}$ is represented versus the coordinate x along the horizontal axis and the time t along the vertical axis. For every escape time no interference fringes are observed. For the color scale see Fig. 6.8.

trajectory of the inset until it hits the slits again. The length of the trajectory is 1.85 m which agrees with the spacing between the peaks in the experimental time spectrum (upper left panel). For $\alpha = 54°$ the reflected part of the wave packet at the slits flies to the upper right corner and returns to the slits. This distance corresponds to half the length of the family of periodic orbits shown in the inset. It equals 1.24 m what again agrees well with the spacing between the peaks in the experimental time spectrum (lower left panel).

The right panels of Fig. 6.17 show the time averaged intensities as function of x. They do not show interference fringes. Figure 6.18 shows the time evolution of the intensity patterns for a wave packet prepared with $\alpha = 54°$. It reveals that even the intensity patterns for a given escape time do not show interference fringes, which is in contrast to the case of a single emitting antenna (cf. Fig. 6.13). An explanation for this behaviour is still lacking.

The same experiment was done in the tilted stadium with an array of 43 antennas placed 10 mm apart from the left boundary as in the rectangular billiard. The distance Δy between adjacent antennas is set again to 7.5 mm[7]. The intensity of the waves escaping through the slits is plotted in the left panels of Fig. 6.19 as function of the time of flight (in length units). The initial angles α are 21°, 41°, and 61° (from up to down). The peak positions show only a weak dependence on the initial angle but their amplitudes vary appreciably except that of the highest peak at about 7.5 m. The distance between successive peaks equals 0.8 m and corresponds to the period of the shortest unstable periodic orbits, i.e. the diffractive bouncing ball orbit with length 0.83 m and the orbit reflected in the circular arc and in the opposite corner (0.92 m) (see insets of the upper left panel of Fig. 6.19). This hints the dominance of these orbits in the patterns. This is also suggested by two further observations. On the one hand the numerical computations of wavefunctions show localization on these orbits. Furthermore the ray tracing simulations presented in Sec. 3 indicate the trapping of particles in the vicinity of the mentioned periodic orbits. All this shows that indeed, these two periodic orbits play an important role in the dynamics of the stadium billiard. The time averaged intensity patterns (right panels of Fig. 6.19) reflect a weak non symmetrical interference structure which depends only slightly on the initial angle.

The very clear periodic structure of the escape times in the rectangular billiard leaves no doubt that the achievement of an initial state with a preferred direction is obtained. Then, in spite of the mentioned open questions, i.e. the weak interference structure in the case of the stadium billiard and the absence of interferences in the case of the rectangular billiard, the method of combining time spectra from different antennas enables the preparation of a wave packet with a well defined direction of propagation. This will be further exploited in the next section.

---

[7] The reason for taking 43 antennas instead of 59 lies on the smaller length of the left billiard´s boundary (370 mm in the stadium versus 475 mm in the rectangle).

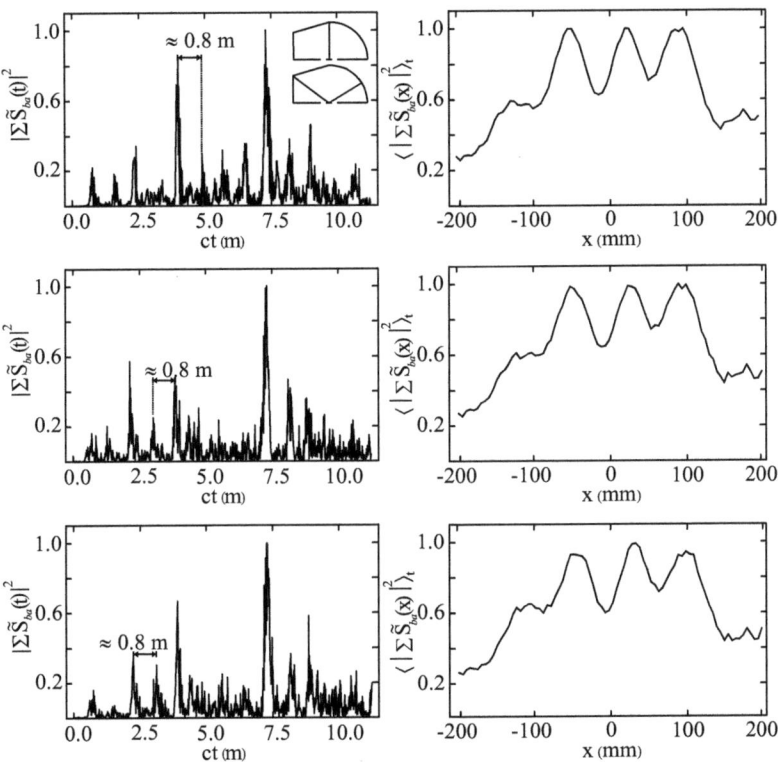

Fig. 6.19: Time spectra (left panels) and time-averaged intensity patterns (right panels) for the elongated wave packet prepared in the stadium billiard with an array of 43 antennas. The initial angles are set to 21°, 41°, and 61° (from up to down).

6.6 Excitation of the billiards with a Gaussian wave packet

6.6.1 Preparation of a Gaussian wave packet

Using the same principle as in the previous section we now try to prepare a wave packet propagating in a well defined direction. Similarly to the numerical simulation of [43] presented in Sec. 6.1, we choose a Gaussian wave packet as initial state,

$$E_z(\vec{r}, t_0 = 0) = \exp\left(-\frac{(\vec{r}-\vec{r}_0)^2}{2\sigma^2}\right) \exp(i\vec{k}(\vec{r}-\vec{r}_0)), \quad (6.18)$$

where $\vec{r}_0$ denotes the central position of the initial wave packet and σ its width.

According to the propagator formalism described in Sec. 6.4.1 an initial state evolves in the following way

$$E_z(\vec{r}, t) \propto \int_\mathcal{G} d^2\vec{r}' \tilde{S}_{ba}(\vec{r}, \vec{r}', t, t_0) \frac{\partial E_z(\vec{r}', t_0)}{\partial t} - \int_\mathcal{G} d^2\vec{r}' \frac{\partial \tilde{S}_{ba}(\vec{r}, \vec{r}', t, t_0)}{\partial t} E_z(\vec{r}', t_0), \quad (6.19)$$

where the propagator $K(\vec{r}, \vec{r}', t, t_0)$ in Eq. (6.11) has been replaced by the Fourier transform of the transmission amplitude $\tilde{S}_{ba}(\vec{r}, \vec{r}', t, t_0)$ between two antennas at positions $\vec{r}'$ and $\vec{r}$. As in Sec. 6.5, the second term in Eq. (6.19) can be set to zero by proper choice of initial conditions, i.e. $E_z(\vec{r}', t_0) = 0$. Since we can obtain $S_{ba}(\vec{r}, \vec{r}', t, t_0)$ experimentally only for discrete antenna positions $\vec{r}_{0i}$, the integral entering Eq. (6.19) has to be subsituted by a sum over discrete points $\vec{r}_{0i}$ in the restricted domain where the antennas are placed. Hence, the time evolution is obtained from the equation

$$E_z(\vec{r}, t) \propto \sum_i \tilde{S}(\vec{r}, \vec{r}_{0i}, t, t_0) \frac{\partial E_z(\vec{r}_{0i}, t_0 = 0)}{\partial t}. \quad (6.20)$$

Inserting the time derivative of Eq. (6.18), the final expression for the constructed wave packet reads

$$E_z(\vec{r}, t) \propto -i\omega \sum_i \exp\left(-\frac{(\vec{r}_{0i}-\vec{r}_0)^2}{2\sigma^2}\right) \exp(i\vec{k}(\vec{r}_{0i}-\vec{r}_0)) \tilde{S}_{ba}(\vec{r}, \vec{r}_{0i}, t, t_0). \quad (6.21)$$

It is proportional to a sum over the measured time spectra weighted by complex coefficients. The initial direction and the width of the wave packet can be varied arbitrarily since $\vec{k} = k(\cos\alpha, \sin\alpha)$ and σ are free parameters. However, the width σ should be chosen larger than the minimal wavelength used in the experiments, $\lambda_{\min} = 15$ mm. On the other hand if it is chosen much larger than the size of the domain of emitting antennas the coefficents would be all approximately 1 and all antennas would equally contribute. Thus σ is chosen of the order of the size of this domain. The time evolution of such a constructed electromagnetic wave packet is used in the experiments for the study of the dependence of the interference pattern on the billiard geometry as well as on the initial angle, and so enable a more direct comparison with the simulation by Prosen and Casati [43].

6.6.2 Experiments with a Gaussian wave packet in free space

As in the case of a single emitting antenna (see Sec. 6.4.2) the propagation of the wave packet constructed according Eq. (6.21) can be tested experimentally in free space. The wave packet is constructed from a squared array of 5×5 positions $\vec{r}_a$ of the emitting antenna and the transmission spectra $S_{ba}(\vec{r}_b, \vec{r}_a, f)$ are measured. For each of the 25 positions $\vec{r}_a$, the emitting antenna hangs from the ceiling of an empty room and the receiving antenna is moved, i.e. $\vec{r}_b$ is varied in the vicinity of $\vec{r}_a$ in a plane perpendicular to the metallic wire of the emitting antenna. The experimental parameters $\Delta x = 5$ mm, $\Delta y = 5$ mm, $f_{\min} = 0.5$ GHz and $f_{\max} = 20$ GHz are taken as those of the measurements described in Sec. 6.4.2. The independent frequency spectra for each emitting antenna position are Fourier transformed into the time domain and finally the time spectra are added up according to Eq. (6.21). The result is shown in Fig. 6.20 for three snapshots of the wave packet propagation. The width of the packet σ is 20 mm and the initial angle α depicted in Fig. 6.21 equals 31°. Thus, the electric field can be obtained as function of the spatial coordinates and time. The wave packet moves into the expected direction with an angle $\alpha = 31°$ and spreads out fastly. The spatial structure of the spreaded wave packet reveals propagation in additional directions. We attribute this effect to the discretization of the source points.

6.6.3 Experiments with a Gaussian wave packet in microwave billiards

For the construction of an initial wave packet with a directional propagation as described in Sec. 6.6.1, a squared array of 25 antennas is drilled in the top plates of the billiards at distances of 7.5 mm as shown in the sketch of the left picture of Fig. 6.21.

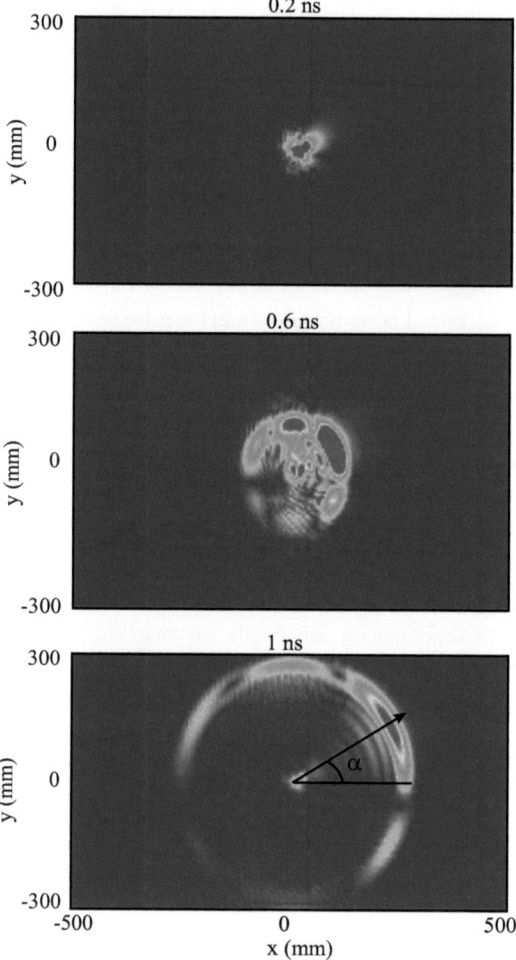

Fig. 6.20: Three snapshots of the propagation in free space of a Gaussian wave packet constructed from 25 emitting antennas. For the color scale see Fig. 6.8.

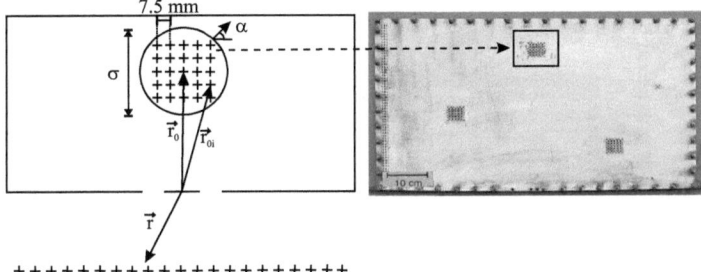

Fig. 6.21: Sketch and photography of the experimental disposition for constructing a wave packet inside the resonators. 25 holes for the antennas are drilled into the top billiard plates.

The right picture is a photography of the top plate of the rectangular billiard. The position of the central antenna of the squared array is located on the line $x=0$ at a distance of 400 mm from the lower edge of the billiards. Similarly to the elongated wave packet construction (see Sec. 6.5), initial angles are chosen such the wave packet initially moves along periodic orbits in the rectangular billiard. The angle α is defined as indicated in Fig. 6.21.

Three cases corresponding to different values of the inital angle α are analyzed in the following. Figure 6.22 shows in the left panels the time spectrum and in the right panels the time-averaged intensity pattern for an initial Gaussian wave packet with parameters $\alpha = 90°$, $\sigma = 30$ mm (top panels), $\alpha = 54°$, $\sigma = 30$ mm (middle panels) and $\alpha = 22°$, $\sigma = 30$ mm (bottom panels) in the rectangular billiard. The case $\alpha = 90°$ corresponds to an initial direction towards the slits, i.e. the wave packet moves along the vertical bouncing ball orbit. Thus it travels hence and forth reflecting at the upper and the lower edges and the escape times exhibit a clear periodicity. The period of the escape times in the left panel agrees well with the length of the bouncing ball orbit (0.95 m). In the top right panel of Fig. 6.22 the time averaged intensity pattern displays interference fringes with visibility of close to 100 % (the visibility was defined in Sec. 6.3).

The initial angle $\alpha = 54°$ corresponds to the periodic orbit shown in the inset of the left middle panel of Fig. 6.22. The peak structure observed in the

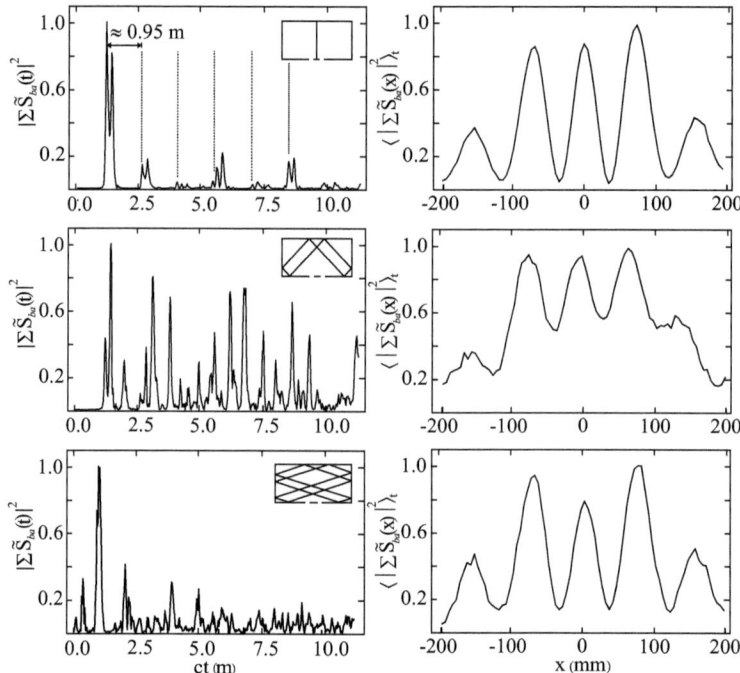

Fig. 6.22: Time spectra (left panels) and time-averaged intensity patterns (right panels) for a constructed wave packet starting with an initial angle of 90° (top), 54° (middle) and 22° (bottom) in the rectangular billiard. The dotted lines in the top left panel indicate the periodicity of the escape times.

middle left panel shows no clear periodicity. We believe that this originates from the spreading of the wave packet in additional directions, cf. Fig. 6.20. The visibility of the interference pattern in the middle right panel of Fig. 6.22 is approximately 50%.

The differences between patterns from the rectangular and the stadium billiard are especially remarkable for initial angles close to 20°. The initial angle $\alpha = 22°$ corresponds to the orbit depicted in the inset of the bottom left panel of Fig. 6.22. In the bottom left panel the peaks appear stochastically, but in the bottom right panel the spatial intensity shows interference fringes. The visibility is 86 %. The presence of a few prominent peaks, especially that at about 1.5 m in the time spectrum might suggest that these peaks dominate the interference pattern. However, the form of the interference pattern is barely modified if the largest peak is removed.

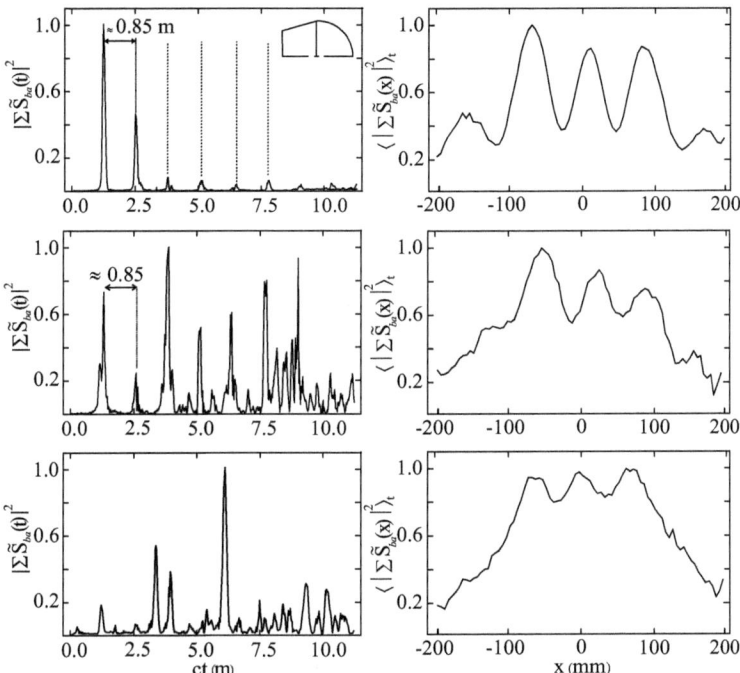

Fig. 6.23: Time spectra (left panels) and time-averaged intensity patterns (right panels) for a constructed wave packet starting with an initial angle of 90° (top), 54° (middle) and 22° (bottom) in the stadium billiard. The dotted lines in the top left panel indicate the periodicity of the escape times.

The same inital angles are chosen in the stadium billiard. The results are shown in Fig. 6.23. The peaks in the top left panel ($\alpha = 90°$) again appear periodically. The period equals the length of the vertical diffractive orbit shown in the inset. The length of this orbit is 0.83 m. In the top right panel the visibility of the interferences has decreased to approximately 60 %. In addition they are slightly non symmetrical. In spite of remainings of interferences for the stadium billiard, the billiard dynamics leads to visible differences in the interference patterns (compare with the top right panel of Fig. 6.22).

The dominant peaks in the middle left panel ($\alpha = 54°$) display a periodic structure corresponding to the period of the vertical diffractive orbit. We again suspect that the wave packet sticks in the vicinity of this periodic orbit. In addition further narrow peaks in between are observed. The visibility of the interference fringes shown in the middle right panel of Fig. 6.23 is moderately lower than the one for the rectangular billiard (see the middle right panel of Fig. 6.22).

In the bottom left panel ($\alpha = 22°$) the escape times exhibit no periodicity. The interferences in the bottom right panel disappear almost completely and the visibility equals 17 %.

Hence the largest discrepancies in the behaviour of the intensity patterns are observed for the initial angle $\alpha = 22°$. A question which inmediately arises is why in the regular case interferences are well-established for certain initial angles but not for others.

6.7 Conclusions

The emission of a wave pulse from a single antenna generates an onmidirectional pulse. This provides no possibility to distinguish the effect of the billiard dynamics (regular or chaotic) on the interference patterns formed by the part of the wave pulse exiting via two slits in the boundary. Unless the emitting antenna lies on the symmetry line defined by the two slits, the portions of the wave pulse which interfere after exiting the slits travel along different paths. Correspondingly, the interference fringes measured along a line at a certain distance from the lower edge of the billiard change their spatial positions with time. Thus, when averaging over time the interferences are unavoidably lost. Hence the choice of an initial momentum (i.e. definite direction) seems to be crucial for the emergence of interference patterns.

A first experiment to control the initial direction of propagation of the constructed wave packet was performed with a vertical array of antennas. The excitation of subsequent antennas with a certain time shift allows the construction of an elongated propagation front. In the rectangular cavity the measured time spectra $|\tilde{S}_{ba}|^2$ reproduce quite well the periodicity of the classical periodic orbits moving in the direction of propagation (at least the shortest ones). However measured intensity patterns yield no interferences for the elongated wave packet. Moreover very weak interferences are formed for the patterns corresponding to the stadium billiard. In addition the time spectrum of the stadium exhibit periodicities corresponding to the vertical diffractive periodic orbit. In order to supress its effect, these experiments have also been performed with absorption material inserted at the corner enclosed by the quarter circle and the tilted wall. Despite this corner is hit by the diffractive orbit, similar results have been obtained. This finding suggests that other orbits reinforced by the focusing property of the circular arc may generate the mentioned periodicity and the remnants of interferences.

A second experiment was carried out with a square array configuration of 25 antennas. The combination of the spectra measured with each of the antennas provides a method to create a directed wave packet. A test of the propagation of an electromagnetic Gaussian wave packet in free space shows a fast spreading after few ns and additional propagation directions. In the stadium billiard remnants of interferences are observed practically for all initial angles although the visibility is in general larger for the rectangular cavity. For angles close to 20° the interferences are almost completely supressed for the stadium billiard whereas they arise with high visibility (about 85 %) for the rectangular one. The underlying mechanism for the formation of interferences is not fully understood since the high sensitivity of the patterns to initial parameters of the wave packet does not allow to pose robust conclusions. Hence the result presented by Casati and Prosen [43] is approximately reproduced only for certain initial directions. It should be remembered that in the simulation of Casati and Prosen the sharp distinction between patterns from billiards with regular and chaotic dynamics was only obtained for certain initial conditions, too.

7 Final considerations

In the present work the decay rate and the spatial structure of the field leaking from microwave billiards opened at the boundary via one or two holes have been investigated. The effect of the dynamics, i.e. regular or chaotic, on the temporal decay behavior and on the wave patterns has been explored. Since every physical measured system is by definition coupled to its environment (by means of the measurement apparatus) and therefore not isolated, the topics covered by this work may be relevant not only for the study of open quantum systems but also for practical applications. The problem of quantum transport in mesoscopical devices like quantum dots and quantum corrals is very closely related to the escape of a quantum particle from a billiard [31]. According to the results shown in Sec. 5, the quantum mechanical temporal behaviour follows that of the corresponding classical one up to about 100 t_H and deviates for longer times due to the influence of a few narrow resonances. This was also found in other publications [35, 69].

The measurements of the decay probability in circle billiards with small pieces of microwave absorber attached to the boundary (see Ref. [94]) instead of a real opening have revealed interesting results. Ideally, absorption material does not reflect microwave radiation. Thus it seems to be very similar to a hole in classical physics. In fact, the structure of the resonant states measured in systems with absorbers and real openings differ in a very considerable way. In measurements with absorbers, the states which are symmetric with respect to the axis defined by the opening are really difficult to observe in the spectra since they are strongly absorbed (see dashed line of Fig. 7.1). Because of this, they do not account for the decay rate. In contrast, in measurements with real openings the waves are diffracted on the edges of the opening. A part is reflected into the cavity and a part leaves the system through the opening. Consequently, the short-lived states can be measured and they contribute to the fast decay (see solid line of Fig. 7.1).

In the stationary regime results of double-slit experiments are in accordance with those obtained from similar experiments with water surface waves in billiards [103] only in the case of regular dynamics, where for both systems the intensity equals the sum of the intensities from two single-slit experiments plus an interference term. For the case of chaotic dynamics, however, the random plane wave model, which has been shown to describe well wavefunction properties of chaotic systems [5], does not describe the experimental observations. In the time domain the crucial question is how exactly the spreading of the initial wave packet affects the resulting interference patterns. In the numerical simulation of [43], i.e. quantum-mechanical situation, the spreading of a Gaussian wave packet is determined by the (quasi-)unitary evolution of the propagator. In the electromagnetic analog examined in this thesis, the construction of a Gaussian wave packet can be achieved only approximately since the number of discrete sources is finite. Moreover, the spreading of the packet has been only tested in free space and for very short times. Since the long-time evolution of the wave packet remains unclear it is difficult to estimate the exact consequences of the rapid spreading on the interferences. However, the appearance of periodicities in the escape times has revealed the sticking of the constructed wave packet on periodic orbits. On the other hand the authors of [43] have recognized that in the case of regular dynamics, not all initial configurations produced interference fringes in the simulation. This is in fact the situation observed in the experiments. Experiments with well directed electromagnetic radiation (like for example beams of laser light confined in two-slit optical cavities) may extend the results presented in this doctoral thesis.

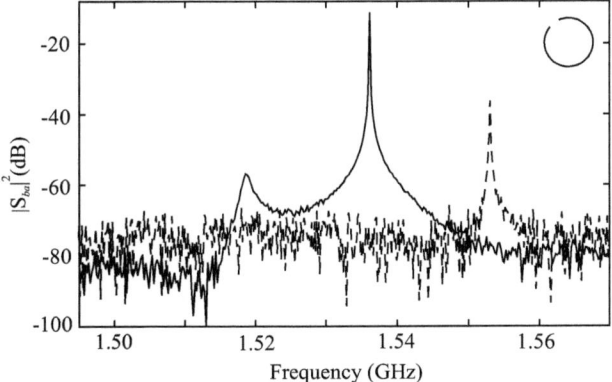

Fig. 7.1: Comparison of the lowest-lying doublet of the circular billard measured in an open billiard with a real opening (solid line) and with an equivalent piece of absorber attached to the boundary (dashed line).

References

[1] H. Poincaré: *Les methodes nouvelles de la mechanique celeste* (Gauthier-Villars, Paris, 1899).

[2] E. Zahar: *Poincare's Philosophy: From Conventionalism to Phenomenology* (Open Court Pub Co., New York, 2001).

[3] H. Bai-Lin: *Chaos* (World Scientific, Signapore, 1984).

[4] H. G. Schuster: *Deterministic Chaos*, 2nd edition (VCH, Weinheim, Germany, 1988).

[5] M. Berry: *Chaotic Behavior of Deterministic Systems* (North Holland, Amsterdam, 1983).

[6] J. Gollub and G. Baker: *Chaotic Dynamics: An Introduction* (Cambridge University Press, Cambridge, 1996).

[7] M.-J.Giannoni, A.Voros, and J.Zinn-Justin: *Chaos and Quantum Physics* (Elsevier Sci. Publ. B.V., Les Houches, 1991).

[8] A. Ben-Mizrachi, I. Procaccia, N. Rosenberg, A. Schmidt, and H. G. Schuster: *Real and Apparent Divergencies in Low-Frequency Spectra of Nonlinear Dynamical Systems*, Phys. Rev. A **31**, 1830 (1985).

[9] K. Briggs: *Simple experiments in chaotic dynamics*, Am. J. Phys. **55**, 1083 (1987).

[10] F. Moon: *Experiments in Chaotic Dynamics* (Springer-Verlag, New York, 1988).

[11] P. Cvitanovic, R. Artuso, R. Mainieri, G. Tanner, G. Vattay, N. Whelan, and A. Wirzba: *Chaos: Classical and Quantum*, http://www.chaosbook.org, 2009.

[12] M. Gutzwiller: *Chaos in Classical and Quantum Mechanics* (Cambridge University Press, Springer, New York, 1990).

[13] O. Bohigas, M. Giannoni, and C. Schmit: *Characterization of chaotic spectra and universality of level fluctuation laws*, Phys. Rev. Lett. **52**, 1 (1984).

[14] H.-J. Stöckmann: *Quantum Chaos: An Introduction* (Cambridge University Press, Cambridge, 1999).

[15] H.-D. Gräf, H. L. Harney, H. Lengeler, C. H. Lewenkopf, C. Rangarchayulu, A. Richter, P. Schardt, and H. A. Weidenmüller: *Distribution of Eigenmodes in a Superconducting Stadium Billiard with Chaotic Dynamics*, Phys. Rev. Lett. **69**, 1296 (1992).

[16] H. Alt, H.-D. Gräf, H. L. Harney, R. Hofferbert, H. Lengeler, A. Richter, P. Schardt, and H. A. Weidenmüller: *Gaussian Orthogonal Ensemble Statistics in a Microwave Stadium Billiard with Chaotic Dynamics: Porter-Thomas Distribution and Algebraic Decay of Time Correlations*, Phys. Rev. Lett. **74**, 62 (1995).

[17] J. Stein and H.-J. Stöckmann: *Microwave Studies of Billiard Green Functions and Propagators*, Phys. Rev. Lett. **75**, 53 (1995).

[18] E. J. Heller: *Quantum localization and the rate of exploration of phase space*, Phys. Rev. A **35**, 1360 (1987).

[19] J. Wilkie and P. Brumer: *Time-Dependent Manifestations of Quantum Chaos*, Phys. Rev. Lett. **67**, 1185 (1991).

[20] S. Tomsovic and E. J. Heller: *Long-time semiclassical dynamics of chaos: The stadium billiard*, Phys. Rev. E **47**, 282 (1993).

[21] P. Leboeuf and G. Iacomelli: *Statistical properties of the time evolution of complex systems.I*, arXiv:cond-mat/9709070, Preprint (1997).

[22] D. V. Savin and V. V. Sokolov: *Quantum versus classical decay laws in open chaotic systems*, Phys. Rev. E **56**, 4911 (1997).

[23] J. Ryu, S. Lee, C. Kim, and Y. Park: *Survival probability time distribution in dielectric cavities*, Phys. Rev. E **73**, 036207 (2006).

[24] V. Milner, J. L. Hanssen, W. C. Campbell, and M. G. Raizen: *Optical billiards for atoms*, Phys. Rev. Lett. **86**, 1514 (2001).

[25] N. Friedmann, A. Kaplan, D. Carasso, and N. Davidson: *Observation of chaotic and regular dynamics in atom-optics billiards*, Phys. Rev. Lett. **86**, 1518 (2001).

[26] W. A. Lin and J. B. Delos: *Order and chaos in semiconductor structures*, Chaos **3**, 1054 (1993).

[27] H. U. Baranger: *Quantum transport and chaos in semiconductor microstructures*, Physica D **83**, 30 (1995).

[28] R. V. Jensen: *Chaotic scattering, unstable periodic orbits and fluctuations in quantum transport*, Chaos **101**, 1 (1991).

[29] B. Huckestein, R. Ketzmerick, and C. H. Lewenkopf: *Quantum transport through ballistic cavities: Soft vs. hard quantum chaos*, Phys. Rev. Lett. **84**, 5504 (2000).

[30] S. Rotter, J. Z. Tang, L. Wirtz, J. Trost, and J. Burgdörfer: *Modular recursive Green's function method for ballistic quantum transport*, Phys. Rev. B **62**, 1950 (2000).

[31] C. D. Schwieters, J. A. Alford, and J. B. Delos: *Semiclassical scattering in a circular semiconductor microstructure*, Phys. Rev. B **54**, 10652 (1996).

[32] M. Puhlmann, H. Schanz, T. Kottos, and T. Geisel: *Quantum decay of an open chaotic system: A semiclassical approach*, Europhys. Lett. **69**, 313 (2004).

[33] C. Stampfer, S. Rotter, J. Burgdörfer, and L. Wirtz: *Pseudopath semiclassical approximation to transport through open quantum billiards: Dyson equation for diffractive scattering*, Phys. Rev. E **72**, 036223 (2005).

[34] T. Blomquist and I. V. Zozoulenko: *Semiclassical transport in a square billiard: Conductance oscillations as probe of coherence length*, Phys. Rev. B **61**, 1724 (2000).

[35] I. V. Zozoulenko and T. Blomquist: *Time-resolved dynamics of electron wave packets in chaotic and regular quantum billiards with leads*, Phys. Rev. B **67**, 085320 (2003).

[36] T. Young: *Course of Lectures on Natural Philosophy and the Mechanical Arts* (Taylor and Walton, London, 1845).

[37] C. Jönsson: *Elektroneninterferenzen an mehreren künstlich hergestellten Feinspalten*, Zeitschrift für Physik **161**, 454 (1961).

[38] A. Tonomura, J. Endo, T. Matsudaa, T. Kawasaki, and H. Ezawa: *Demonstration of single-electron build up of an interference pattern*, Am. J. Phys. **57**, 117 (1989).

[39] A. Zeilinger, R. Gähler, C. Shull, W. Treimer, and W. Mampe: *Single- and double-slit diffraction of neutrons*, Rev. Mod. Phys. **60**, 1067 (1988).

[40] O. Carnal and J. Mlynek: *Young's double-slit experiment with atoms: A simple atom interferometer*, Phys. Rev. Lett. **66**, 2689 (1991).

[41] M. Noel and C. Stroud: *Young's Double-Slit Interferometry within an Atom*, Phys. Rev. Lett. **75**, 1252 (1995).

[42] M. Arndt, O. Nairz, J. Vos-Andreae, C. Keller, G. van der Zouw, and A. Zeilinger: *Wave-particle duality of C60 molecules*, Nature **401**, 680 (1999).

[43] G. Casati and T. Prosen: *Quantum chaos and the double-slit experiment*, Phys. Rev. A **72**, 032111 (2005).

[44] L. Bunimovich: *On the ergodic properties of nowhere dispersing billiards*, Comm. Math. Phys. **65**, 295 (1979).

[45] H. Primack and U. Smilansky: *Quantal consequences of perturbations which destroy structurally unstable orbits in chaotic billiards*, J. Phys. A: Math. Gen. **27**, 4439 (1994).

[46] T. Friedrich: *Eigenschaften von Pilzbillards und Korrelationsfunktionen von Streumatrixelementen in Mikrowellenresonatoren*, Dissertation D17, TU Darmstadt, 2007.

[47] C. Cohen-Tannoudji: *Quantum Mechanics* (John Wiley and Sons, New York, 1977).

[48] J. D. Jackson: *Classical Electrodynamics* (John Wiley and Sons, Inc., New York, 1999).

[49] W. Bauer and G. Bertsch: *Decay of ordered and chaotic systems*, Phys. Rev. Lett. **65**, 2213 (1990).

[50] H. Alt, H.-D. Gräf, H. Harney, R. Hofferbert, H. Rehfeld, A. Richter, and P. Schardt: *Decay of classical chaotic systems: The case of the Bunimovich stadium*, Phys. Rev. E **53**, 2217 (1996).

[51] A. J. Fendrik, A. M. F. Rivas, and M. J. Sánchez: *Decay of quasibounded classical Hamiltonian systems and their internal dynamics*, Phys. Rev. E **50**, 1948 (1994).

[52] A. J. Fendrik and M. J. Sánchez: *Decay of quasibounded classical Hamiltonian systems populated by scattering experiments*, J. Phys. A: Math. Gen. **28**, 4235 (1995).

[53] E. Vicentini and V. B. Kokshenev: *On survival dynamics of classical systems. Non-chaotic open billiards*, Physica A **295**, 391 (2001).

[54] T. Tudorovskiy, R. Höhmann, U. Kuhl, and H.-J. Stöckmann: *On the theory of cavities with point-like perturbations. Part I: General theory*, J. Phys. A: Math. Theor. **41**, 275101 (2008).

[55] M. T. Gelevarzi: *Gekoppelte Mikrowellenbillards unterschiedlicher Größe*, Diplomarbeit, TU Darmstadt, 2006 (unpublished).

[56] R. Balian and C. Bloch: *Distribution of Eigenfrequencies for the Wave Equation in a Finite Domain. II. Electromagnetic Fields. Riemannian Spaces*, Ann. Phys. (NY) **64**, 271 (1970).

[57] H. Alt, P. von Brentano, H.-D. Gräf, R. Hofferbert, M. Philipp, H. Rehfeld, A. Richter, and P. Schardt: *Precision test of the Breit-Wigner formula on resonances in a superconducting microwave cavity*, Phys. Lett. B **366**, 7 (1996).

[58] F. Beck, C. Dembowski, A. Heine, and A. Richter: *R-matrix theory of driven electromagnetic cavities*, Phys. Rev. E **67**, 066208 (2003).

[59] H. Padamsee, J. Knobloch, and T. Hays: *RF Superconductivity for Accelerators* (John Wiley and Sons, Inc., New York, 1998).

[60] H. Alt: *Gekoppelte supraleitende Mikrowellenbillards als Modellsystem für Symmetriebrechung*, Dissertation D17, TU Darmstadt, 1998.

[61] S. Albeverio and F. Haake: *S-matrix, resonances, and wave functions for transport through billiards with leads*, J. Math. Phys. **37**, 4888 (1996).

[62] C. Mahaux and H. A. Weidenmüller: *Shell-Model Approach to Nuclear Reactions* (North-Holland Publishing Company, Amsterdam-London, 1968).

[63] I. Rotter: *Dynamics of quantum systems*, Phys. Rev. E **64**, 036213 (2001).

[64] Y. V. Fyodorov and H.-J. Sommers: *Statistics of resonance poles, phase shifts and time delays in quantum chaotic scattering: Random matrix approach for systems with broken time-reversal invariance*, J.Math.Phys. **38**, 1918 (1997).

[65] F.-M. Dittes: *The decay of quantum systems with a small number of open channels*, Phys. Rep. **339**, 215 (2000).

[66] A. M. Yaglom: *An Introduction to the Theory of Stationary Random Functions* (Prentice Hall, Englewood Cliffs, New Jersey, 1962).

[67] C. B. Chiu, E. C. G. Sudarshan, and B. Misra: *Time evolution of unstable quantum systems and a resolution of Zeno's paradox*, Phys. Rev. D **16**, 520 (1977).

[68] K. Unnikrishnan: *Short- and long-time decay laws and the energy distribution of a decaying state*, Phys. Rev. A **60**, 41 (1999).

[69] J. A. Hart, T. M. Antonsen, and E. Ott: *Scattering a pulse from a chaotic cavity: Transitioning from algebraic to exponential decay*, Phys. Rev. E **79**, 016208 (2009).

[70] J. J. M. Verbaarschot, H. A. Weidenmüller, and M. R. Zirnbauer: *Grassmann integration in stochastic quantum physics: The case of compound-nucleus scattering*, Phys. Rep. **129**, 367 (1985).

[71] V. V. Sokolov: *Decay Rates Statistics of Unstable Classically Chaotic Systems*, in *Nuclei and mesoscopic physics*, **995**, AIP Conference Proceedings, Ed.: P. Danielewicz, P. Piecuch, and V. Zelevinsky (Melville, New York, 2008), p. 85.

[72] T. Kottos: *Statistics of resonances and delay times in random media: beyond random matrix theory*, J. Phys. A: Math.Gen **38**, 10761 (2005).

[73] Y. V. Fyodorov and H.-J. Sommers: *Statistics of S-matrix poles in few-channel chaotic scattering: crossover from isolated to overlapping resonances*, JETPL **63**, 1026 (1995).

[74] C. Dembowski, B. Dietz, T. Friedrich, H.-D. Gräf, H. L. Harney, A. Heine, M. Miski-Oglu, and A. Richter: *Distribution of resonance strengths in microwave billiards of mixed and chaotic dynamics*, Phys. Rev. E **71**, 046202 (2005).

[75] P. Šeba, F. Haake, M. Kus, M. Barth, U. Kuhl, and H.-J. Stöckmann: *Distribution of the wave function inside chaotic partially open systems*, Phys. Rev. E **56**, 2680 (1997).

[76] M. Miski-Oglu: *Superscars und Nodal Domain Statistik im symmetrischen Barrierenbillard*, Dissertation D17, TU Darmstadt, 2007.

[77] L. C. Maier and J. C. Slater: *Field strength measurements in resonant cavities*, J. Appl. Phys. **23**, 68 (1952).

[78] E. Bogomolny, B., T. Friedrich, M. Miski-Oglu, A. Richter, F. Schäfer, and C. Schmit: *First experimental observation of superscars in a pseudointegrable barrier billiard*, Phys. Rev. Lett. **97**, 254102 (1997).

[79] R. G. Nazmitdinov, K. N. Pichugin, I. Rotter, and P. Šeba: *Whispering gallery modes in open quantum billiards*, Phys. Rev. E **64**, 056214 (2001).

[80] R. G. Nazmitdinov, K. N. Pichugin, I. Rotter, and P. Šeba: *Conductance of open quantum billiards and classical trajectories*, Phys. Rev. B **66**, 085322 (2002).

[81] T. Gorin, D. Martinez, and H. Schomerus: *Long-time signatures of short-time dynamics in decaying quantum-chaotic systems*, Phys. Rev. E **75**, 016217 (2007).

[82] E. Doron, U. Smilansky, and A. Frenkel: *Experimental Demonstration of Chaotic Scattering of Microwaves*, Phys. Rev. Lett. **65**, 3072 (1990).

[83] E. Doron, U. Smilansky, and A. Frenkel: *Chaotic scattering and transmission fluctuations*, Physica D **50**, 367 (1991).

[84] D. V. Savin and H.-J. Sommers: *Delay times and reflection in chaotic cavities with absorption*, Phys. Rev. E **68**, 036211 (2003).

[85] F. Schäfer: *Time-Reversal Symmetry Breaking in Quantum Billiards*, Dissertation D17, TU Darmstadt, 2009.

[86] B. Dietz, T. Friedrich, H. L. Harney, M. Miski-Oglu, A. Richter, F. Schäfer, and H. A. Weidenmüller: *Chaotic scattering in the regime of weakly overlapping resonances*, Phys. Rev. E **78**, 055204 (2008).

[87] D. V. Savin and H.-J. Sommers: *Distribution of reflection eigenvalues in many-channel chaotic cavities*, Phys. Rev. E **69**, 035201 (2004).

[88] E. Persson, T. Gorin, and I. Rotter: *Resonance trapping and saturation of decay widths*, Phys. Rev. E **58**, 1334 (1998).

[89] E. Persson, T. Gorin, and I. Rotter: *Decay rates of resonance states at high level density*, Phys. Rev. E **54**, 3339 (1996).

[90] E. Persson, K. Pichugin, I. Rotter, and P. Šeba: *Interfering resonances in a quantum billiard*, Phys. Rev. E **58**, 8001 (1998).

[91] H.-J. Stöckmann, E. Persson, Y.-H. Kim, M. Barth, U. Kuhl, and I. Rotter: *Observation of resonance trapping in an open microwave cavity*, Phys. Rev. E **65**, 066211 (2002).

[92] P. Šeba, I. Rotter, M. Müller, E. Persson, and K. Pichugin: *Collective modes in an open microwave billiard*, Phys. Rev. E **61**, 66 (2000).

[93] M. Brack and R. Bhaduri: *Semiclassical Physics* (Addison-Wesley, New York, 1997).

[94] P. Oria-Iriarte: *Vorbereitende Experimente zum Test der Riemannschen Hypothese in offenen Kreisbillards*, Diplomarbeit, TU Darmstadt, 2006 (unpublished).

[95] E. G. Altmann, T. Friedrich, A. E. Motter, H. Kantz, and A. Richter: *Prevalence of marginally unstable periodic orbits in chaotic billiards*, Phys. Rev. E **77**, 016205 (2008).

[96] J. von Fraunhofer: *Joseph von Fraunhofer's gesammelte Schriften* (Verlag der K. Akademie, München, 1888).

[97] E. Hecht: *Optics*, 3rd edition (Addison-Wesley, Massachusetts, 2000).

[98] J. Keller: *Geometrical Theory of Diffraction*, JOSA **52**, 116 (1962).

[99] P. Morse and P. Rubenstein: *The Diffraction of Waves by Ribbons and by Slits*, Phys. Rev. **54**, 895 (1938).

[100] T. Prosen: 2009, private communication.

[101] M. Miski-Oglu: 2008, private communication.

[102] *Emerson and Cuming microwave products*, www.eccosorb.com, 2008.

[103] Y. Tang, Y. Shen, J. Yang, X. Liu, J. Zi, and B. Li: *Experimental evidence of wave chaos from a double slit experiment with water surface waves*, Phys. Rev. E **78**, 047201 (2008).

[104] A. Hibbs and R. Feynman: *Quantum Mechanics and Path Integrals* (McGraw-Hill, New York, 1965).

[105] P. M. Morse and H. Feshbach: *Methods of Theoretical Physics* (McGraw-Hill, New York, 1953).

Acknowledgments

I would like to thank Professor Dr. Dr. h.c. mult. Achim Richter for the confidence he has always had in my work and also in me since I came to Darmstadt as an exchange student. His scientific interest and rigor as well as his firm commitment for the young people will be a model for the rest of my life. I am also very grateful to Professor Dr. Tomaž Prosen for his toughtful disposition when collaborating with experimenters and his impressive talent as a theoretical physicist.

I thank Professor Dr. Hanns Ludwig Harney, Dr. Eugene Bogomolny, Professor Dr. Thomas Seligman, Dr. Carlos Viviescas, Professor Dr. Frank Dittes, Dr. Martha Gutierrez, Dr. Oriol Bohigas and Professor Dr. Hans-Jürgen Stöckmann for very stimulating discussions.

For the technical support during the measurements at superconducting conditions I thank Dr. Ralf Eichhorn and his group of the S-DALINAC. Dipl.-Ing. Stefan Ziemann has always made sure that nothing was missing during my experiments in the anechoic chamber of the Hans-Busch institute of the TU Darmstadt.

I would very especially like to say thanks to the members of the Darmstädter quantum chaos group since honestly this work would have been simply impossible without the help, advice and collaboration of each of them: Dr. Maksym Miski-Oglu for his great ideas, experimental skills and good mood which have always illuminated this work. He was always there when it was neccessary. Dr. Barbara Dietz for the perfect supervision and the corrections and efforts in the theoretical development of this work. Dr. Florian Schäfer for being the most diligent spelling checker, his computer talents and his always friendly disposition and attitude. Dipl.-Phys. Stefan Bittner for his generous help during the hours that we have spent together in the laboratory and the interest that he has always shown in my results.

Also an affectionate "Thank You" to the former Alumni of the quantum chaos group, Dr. Thomas Friedrich and Dipl.-Phys. Majid Taheri, mainly for the instruction of the superconducting measurements and the funny times we have spent.

For the financial support I will thank the Fundación La Caixa and the DAAD. This work was supported by DFG within SFB 634 and the Graduiertenkolleg 410.

Lebenslauf

Pedro Oria Iriarte

12. Dezember 1982	Geboren in Pamplona (Spanien)
1988–2000	Besuch der Schule Sta. Maria la Real, Pamplona
Juni 2000	Abitur in Spanien
2000–2004	Studium der Physik an der Universidad de Salamanca, Spanien
2004–2005	Erasmus-Student an der Technischen Universität Darmstadt
September 2005	Abschluss des Studiums
seit Oktober 2006	Doktorand am Institut für Kernphysik der Technischen Universität Darmstadt
2006–2008	Promotionsstipendium der La Caixa Stiftung - DAAD
2008–2009	Promotionsstipendium des DAAD
Seit 2009	Wissenschaftlicher Mitarbeiter im SFB 634 der DFG

Eidesstattliche Erklärung:
Hiermit erkläre ich, dass ich die vorliegende Dissertation selbständig verfasst, keine anderen als die angegebenen Hilfsmittel verwendet und bisher noch keinen Promotionsversuch unternommen habe.

Darmstadt, im Februar 2010

I want morebooks!

Buy your books fast and straightforward online - at one of world's fastest growing online book stores! Environmentally sound due to Print-on-Demand technologies.

Buy your books online at
www.morebooks.shop

Kaufen Sie Ihre Bücher schnell und unkompliziert online – auf einer der am schnellsten wachsenden Buchhandelsplattformen weltweit! Dank Print-On-Demand umwelt- und ressourcenschonend produziert.

Bücher schneller online kaufen
www.morebooks.shop

KS OmniScriptum Publishing
Brivibas gatve 197
LV-1039 Riga, Latvia
Telefax: +371 686 204 55

info@omniscriptum.com
www.omniscriptum.com

Printed by Books on Demand GmbH, Norderstedt / Germany